STEREOCHEMISTRY IN ORGANIC COMPOUNDS

STEREOCHEMISTRY IN ORGANIC COMPOUNDS

M. G. ARORA

ANMOL PUBLICATIONS PVT. LTD.
NEW DELHI-110 002 (INDIA)

ANMOL PUBLICATIONS PVT. LTD.
4374/4B, Ansari Road, Daryaganj
New Delhi-110 002

Stereochemistry in Organic Compounds

First Edition 1997
Reprint, 2002
ISBN 81-7488-559-5

PRINTED IN INDIA

Published by J. L. Kumar for Anmol Publications Pvt. Ltd., New Delhi-110 002 and Printed at Mehra Offset Press, Delhi.

Preface

That branch of Chemistry which deals with three-dimensional structure of molecules is known as 'stereochemistry'. Since many of the properties and reactions of organic compounds depend upon the spatial arrangement of atoms and groups. For proper understanding of this the knowledge of stereochemistry is essential. Stereoisomerism, which arises when molecules differ only in the arrangement of their atoms or groups in space is elaborated discussed. Also the different flexible shapes (conformations) which molecules can adopt are explained in this book. Written in simple language and explained with suitable diagrams, the book will prove most useful.

The book will prove immensely useful and informative for the students and researchers in the field. This will also serve as textbook as well as reference book for the students concerned.

To make the book more informative and useful suggestions from our learned fellow teachers are welcome.

M. G. Arora

Stereochemistry in Organic Compounds

Meaning of Stereochemistry

Stereochemistry is the branch of chemistry which deals with the three-dimensional structures of molecues. A knowledge of stereochemistry is essential since many of the properties and reactions of organic compounds depend upon the spatial arrangement of atoms and groups. The first part of the chapter deals with *stereoisomerism*, which arises when molecules differ only in the arrangement of their atoms or groups in space. Also, chapter deals with the different flexible shapes *(conformations)* which molecules can adopt.

The stereoisomerism depends upon the existence of different arrangements of atoms or groups in space. It is convenient to divide stereoisomerism into two types, geometrical and optical, and these are dealt with in the sections which follow. The stereoisomerism can be understood only in three dimensional terms. For example, tartaric acid has been known for centuries, being readily obtained from a crystalline deposit which separated in wine vats in the form of a potassium salt. About 1819 a second acid obtained during the crystallisation of the above potassium salt was called racemic acid *(racemus*, bunch of grapes). The two acids have the same chemical properties and both can be represented as HOOCCH(OH)CH(OH) COOH but they differ in such physical properties as their crystalline form, melting point, and water-solubility. In 1853 Pasteur made yet another compound (mesotartaric acid) differing from the others but having the same structure when represented in the linear form shown above.

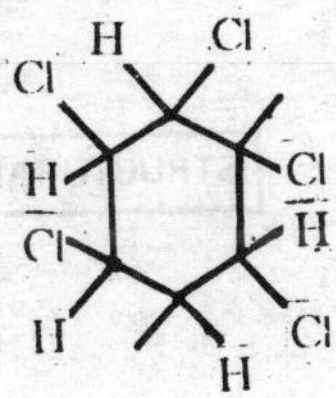

Under appropriate conditions benzene can be chlorinated to give benzene haxachloride. This compound exists in several isomeric forms but only one isomer shows the marked insecticidal action which makes this compound important. Many examples are now known of pharmaceuticals, flavours, and insect attractants, where only one isomer produces the desired effect.

Isomerism

It is common to find that a given molecular formula applies to several quite different organic compounds. For example, the three compounds

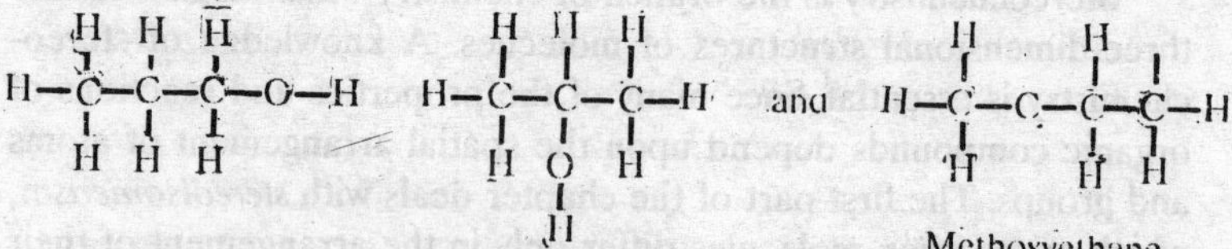

propan-1-ol propan-2-ol, Methoxyethane (or ethyl methyl ether)

all have the molecular formula C_3H_6O, but differ in the arrangement of the atoms in their molecules : they also have different chemical and physical properties. These compounds are said to be *isomeric*.

Types of Isomerism

The two main classes of isomerism are **structural isomerism** and **stereoisomerism.** In structural isomerism the isomers have the same molecular formula but differ in structure, that is, in the order in which the different atoms are linked together in the molecule. In stereoisomerism the isomers have the same molecular formula and the same structure, but differ in the spatial arrangement of the groups.

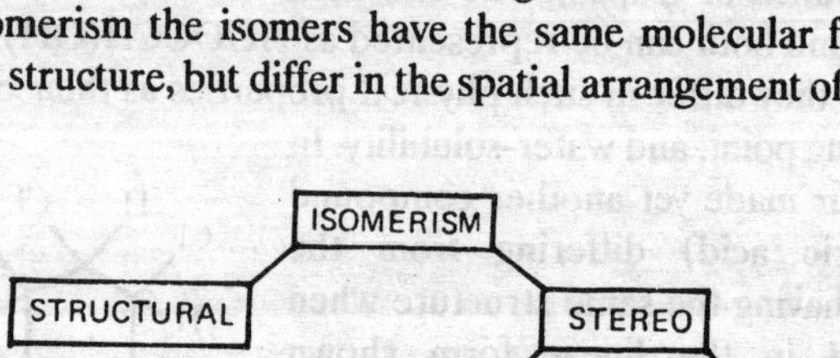

Fig. 1. Types of isomerism encountered in oganic chemistry.

Structural Isomerism

(i) **Chain (or nuclear) isomerism.** Here the isomers differ in the arrangement of the carbon atoms, for example :

$CH_3-CH_2-CH_2-CH_3$ and $CH_3-CH(CH_3)-CH_3$

butane   2-methylpropane

(ii) **Position isomerism.** Positional isomers have the same carbon skeleton and belong to the same homologous series, but differ in the position of the substituent group, and hence usually in their chemical properties; for example :

$CH_3-CH_2-CH_2-Cl$ and $CH_3-CH(Cl)-CH_3$

1-chloropropane   2-chloropropane

Position isomerism is important in cyclic compounds; for example, there are three different isomers of disubstituted benzene, $C_6H_4X_2$. The carbon atoms are numbered to indicate the positions of the substituents.

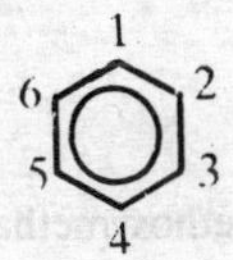

The three isomers of dibromobenzene are :

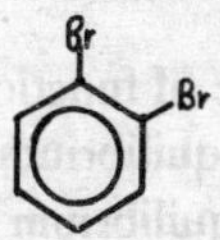

1,2-dibromobenzene

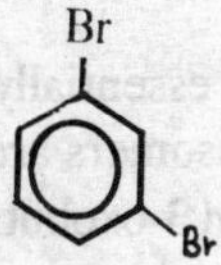

1,3-dibromobenzene

1,4-dibromobenzene

(iii) **Functional-group isomerism.** Functional-group isomers have the same molecular formula but belong to different homologous series, that is, they differ in the nature of the functional group. This type of isomerism is exhibited by many classes of organic ompounds; for instance :

1. Alcohols and ethers, e.g., CH_3CH_2OH and CH_3OCH_3

ethanol   methoxymethane

2. Aldehydes and ketones, e.g. 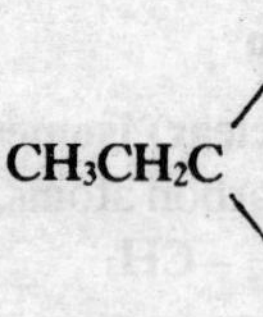 and 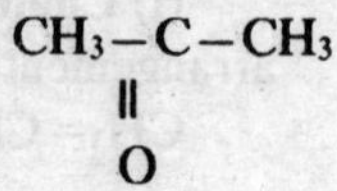

propanal propanone

3. Carboxylic acids and esters, e.g.

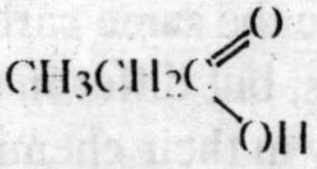

propanoic acid

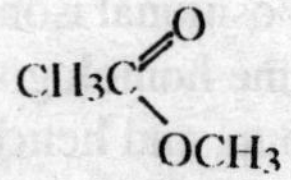 and 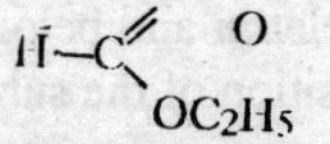

methyl ethanoate ethyl methanoate

(iv) **Metamerism.** Isomers belonging to the same homologous series are known as *metamers*. For example :

1. Ethers, e.g. $C_2H_5OC_2H_5$, $CH_3OCH_2CH_2CH_3$

ethoxyethane methoxypropane

and

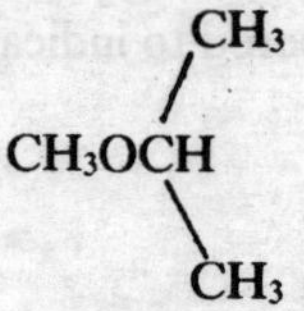

1,1-dimethylmethoxymethane

2. Esters, e.g. $HCO_2C_2H_5$ and $CH_3CO_2CH_3$

ethyl methanoate methyl ethanoate

(v) **Tautomerism** is essentially a special type of functional-group isomerism in which the isomers are in dynamic equilibrium with each other. For example, ethyl 3-oxobutanoate is an equilibrium mixture of two forms :

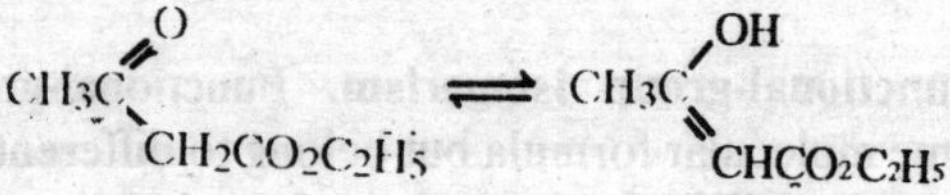

keto form enal form,

At room temperature the mixture contains approximately 93 per cent of the keto- and 7 per cent of the enol forms. The compound shows characteristic reactions of both the carbonyl (>C=O) and the hydroxy-groups.

Tautomerism is caused by the wanderings of a labile hydrogen atom between two polyvalent atoms, carbon and nitrogen.

Some other examples of tautomerism are :

(i) 

Nitroalkane Isonitroalkane

(Resulting from the wanderings of the labile hydrogen atom between the two polyvalent atoms, carbon and oxygen).

(ii) Cyanic acid also exhibits tautomerism and occurs as

H–O–C≡N	O=C=N–H
Cyanic acid gives rise to cyanates	Isocyanic acid gives rise to isocyanates

The labile hydrogen atom oscillates between the polyvalent atoms – oxygen and nitrogen.

Different between Tautomerism and Resonance. Tautomerism should be carefully distinguished from resonance as there is no point of similarity between the two. Some of the important points of difference between them can be enumerated as under :

(i) Tautomers are definite compounds whereas canonical forms of a resonance hybrid are only hypothetical and do not actually exist.

(ii) Tautomerism involves the migration of a labile hydrogen atom between two polyvalent atoms. Thus tautomers differ in atomic arrangement. On the other hand, contributing structures of a resonance hybrid have identical atomic arrangement. These differ only with respect to placement of electron on different atoms.

(iii) Tautomers are in dynamic equilibrium with each other and are connected by sign between them. Since canonical forms of a resonance hybrid are only hypothetical, question of existence of dynamic equilibrium between them does not arise. Resonance is

indicated by a double headed arrow (↔) between two contributing structures.

(iv) One of the requisites for resonance to take place is that the molecule must be planar whereas there is no such requisite for tautomerism.

(v) Resonance is accompanied by an increase in bond length of a double bond and shortening of a single bond length while tautomerism does not involve any change in bond lengths.

(vi) Resonance results in stabilisation of the molecule by lowering its internal energy. There is no such effect in the case of tautomerism.

Stereoisomerism

Stereoisomerism is of two kinds : *geometrical* and *optical isomerism*

(i) **Geometrical isomerism.** Geometrical isomers are a result of hindrance to rotation of a covalent bond, for example, by the presence of a double bond. There are, for istance, two isomeric butenedioic acids :

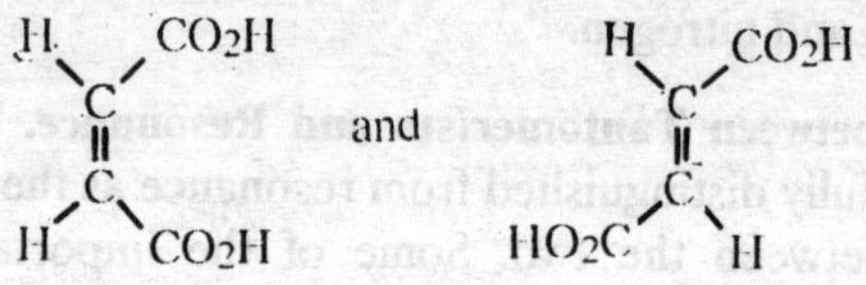

cis-butedioic acid m.p. 130°   *trans*- butenedioic acid m.p. 286°

Both compounds have the same structure; both consist of two carbon atoms, each linked to a hydrogen atom and a carboxy-group, joined together by a double bond. However, they have different chemical and physical properties as a result of the different spatial arrangements of their groups. The prefix *cis*- refers to the positioning of the two carboxy-groups on the same side of the plane of the double bond; in the *trans*-isomer the two carboxy-groups lie on opposite sides.

(ii) **Optical isomerism.** Optical isomers have identical structural formulae and for the most part idential chemical properties but, as a result of the different spatial arrangements of the atoms in their molecules, they differ in their optical properties.

Optical and geometrical isomerism are discussed in detail in this Unit.

Enantiomers and Diastereoisomers

Two or more compounds whose structures differ only in the three-dimensional arrangement of their constitutent atoms are said to be stereoisomers and within this group there are two subgroups : enantiomers and diastereoisomers. *Two stereoisomers which differ only in their ability to rotate plane polarised light in an equal and opposite direction are enantiomers. All other stereoisomeric compounds, whether are able to rotate plane polarised light or not, are diastereoisomers. Enantiomeric pairs are mirror images, diastereoisomers are not.*

This is a wider, but more satisfactory, classification of stereoisomers than was used in the past. Formerly the terms enantiomer and diastereoisomer were applied only to optically active (*asymmetric, chiral*) compounds and *geometric (cis trans)* isomers were not considered as diastereoisomers.

This classification is summarised in the block diagram which is to be interpreted as follows :

(i) stereoisomers are of two distinct kinds : enantiomers and diastereoisomers.

(ii) enantiomers result from chirality only : diastereoisomers result from chirality or *cis-trans*-isomerism.

(iii) chiral systems may be enantiomeric or diastereoisomeric : *cis-trans*-isomers are diastereoisomeric and never enantiomeric (unless they also contain a chiral system).

<table>
<tr><td colspan="3">stereisomers</td></tr>
<tr><td>enantiomers</td><td colspan="2">diastereosiomers</td></tr>
<tr><td colspan="2">chirality</td><td>cis-trans-isomerism</td></tr>
</table>

The Sequence Rules

It will be necessary to designate the *order* of groups (ligands) attached to a particular atom. This is done with the help of so-called Sequence Rules. These ideas owe mainly to Cahn and Ingold (1951) and to Cahn, Ingold, and Prelog (1956, 1966).

It is not sufficient that students *understand* these rules, they must be able to *use* them.

```
    C   P   C
    |   |   |
B - A - X - A - B
    |   |   |
    D   Q   E
```

The Sequence Rules allow us to put the various ligands attached to a particular atom (such as X) in an agreed order of precedence by comparing the ligands according to agreed rules (the Sequence Rules). Comparisons are made between the *atoms* attached to X (P, Q, A, and A in the above figure). If this is not sufficient to settle the order of precedence then the atoms attached to one of the equivalent atoms (B, C, and D in the structure shown) are compared in hierarchical order with the atoms attached to the other equivalent atom also in hierarchical order (B, C, and E). If this does not settle the issue then further sets of atoms come under consideration until a decision can be made.

The rules we shall use state (i) that ligands (as defined above) are arranged in order of *decreasing atomic number*, (ii) that isotopes (when present) are arranged in *decreasing mass number*, and (iii) that when required a lone pair of electrons is placed below hydrogen (^1H) in the precedence order. These rules are applied successively as far as is necessary to reach a decision on priorities, i.e. (ii) only if (i) does not give the answer, and so on. The only other thing we need to know is how to treat unsaturated systems but first let us test our understanding of these simple rules.

Ligands containing unsaturation have to be handled with the help of what are known as "ghost" or "phantom" atoms which are always shown in brackets. Before application of the Sequence Rules the system – A = B, for example, is rewritten as a single-bond structure in which A, instead of being doubly bonded to one atom of B, is singly bonded to two atoms of B. Similarly B must be considered to be singly bonded to two atoms of A so that :

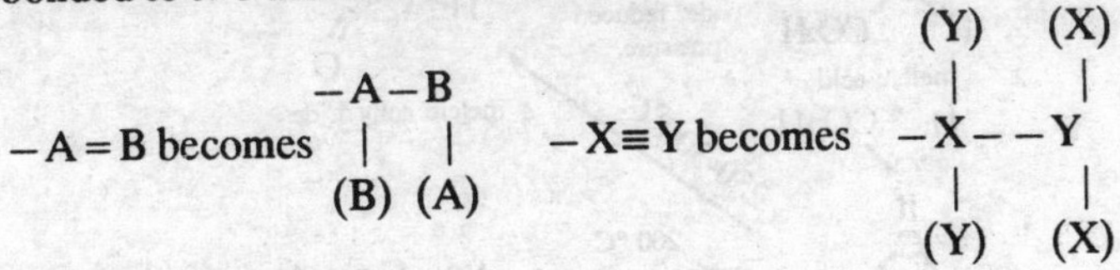

The revised structures contain no multiple bonds and instead a number of ghost atoms.

Geometrical Isomerism

A carbon-carbon double bond introduces a considerable degree of rigidity into a molecule. Thus the relatively free rotation which is possible about a single bond does not occur in the case of a double bond. Rotation of one carbon atom relative to the other requires breaking and reforming the π bond which can occur only under vigorous conditions such as high temperature or irradiation with ultraviolet light.

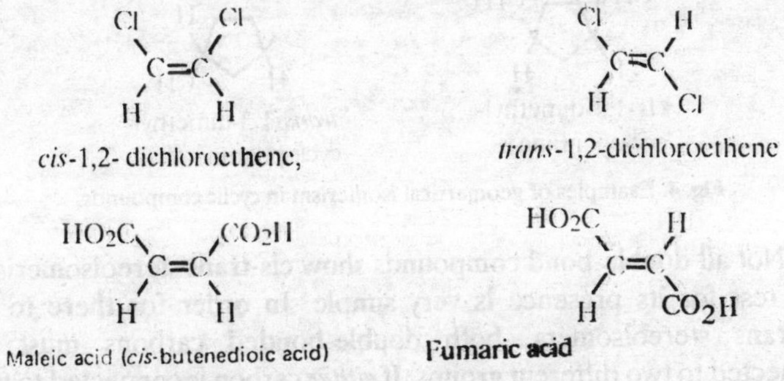

Fig. 2. Examples of geometrical isomerism.

If the two groups attached to each carbon atom of a double bond are different, two isomers exist (Fig. 2). The arrangement of groups about the double bond is referred to as the *configuration* of the molecule. The isomers are often called *cis* and *trans* isomers (in some cases a more general system of nomenclature must be used).

Geometrical isomers usually have different physical and chemical properties. For example, maleic acid (m.p. 130°C) readily forms a

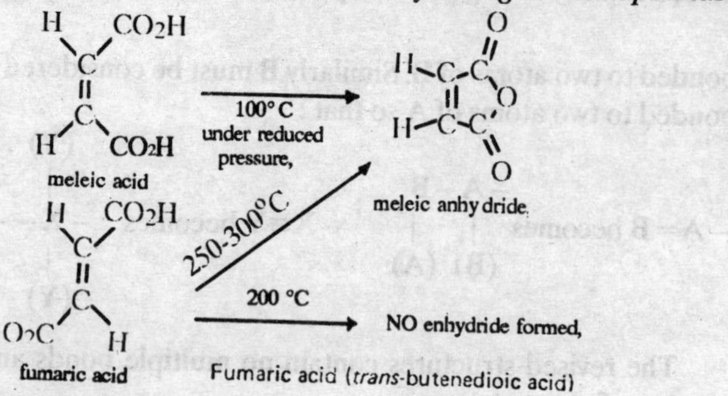

Fig. 3. Differences in the reactions ot geometrical isomers.

cyclic anhydride on heating (Fig. 3), whereas fumaric acid (m.p. 287°C) does not do so unless heated to a very high temperature, when *maleic* anhydride is formed.

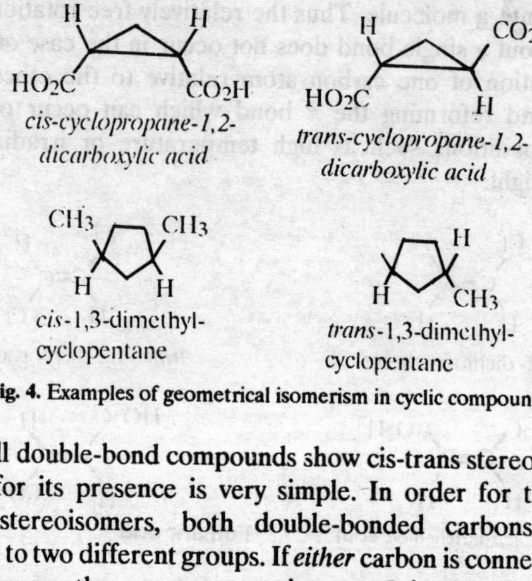

Fig. 4. Examples of geometrical isomerism in cyclic compounds.

Not all double-bond compounds show cis-trans stereoisomerism. The test for its presence is very simple. In order for there to be cis-trans stereoisomers, both double-bonded carbons must be connected to two different groups. If *either* carbon is connected to two identical groups, there are no stereoisomers. It is easy to see why :

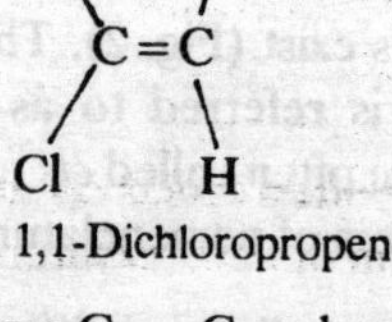

1,1-Dichloropropene

In this case, one of the two C = C carbons is connected to two Cl groups. Thus, if we interchange the groups on either carbon, we get the same compound and not a different one. On the other hand, when both carbons of the double bond are attached to two different groups, we do have cis-trans isomers.

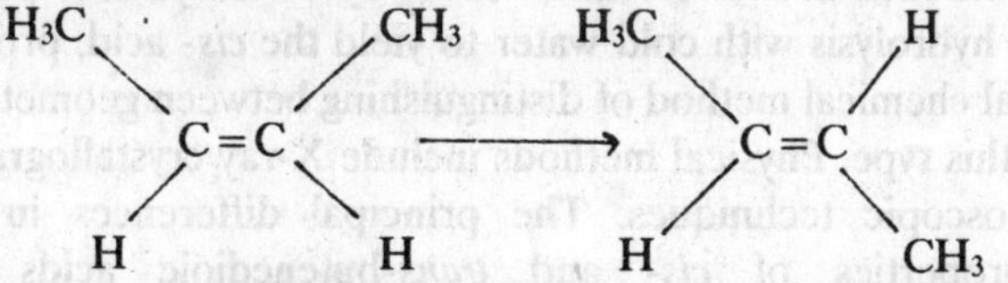

cis-2-Butene *trans*-2-Butene

cis- and _trans_-Butenedioic Acids

Geometrical isomerism occurs in compounds in which free rotation about a particular covalent bond is prevented by the presence of a double bond or of ring structures, or because of steric factors. The best-known example is *butenedioic acid*, which exists in two isomeric forms in which the carbonyl groups lie respectively on the same side and on opposite sides of the doubel bond in the planar molecule :

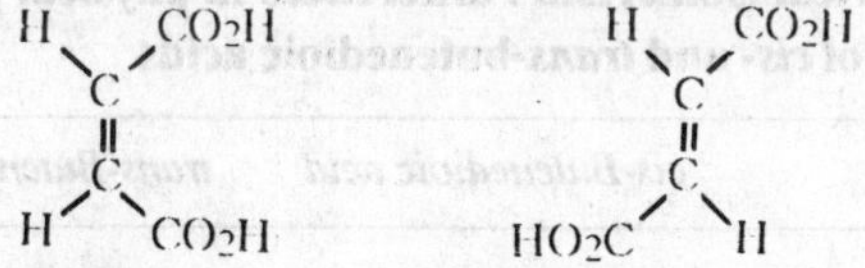

cis-butenedioic acid (maleic acid) *trans*-butenedioic acid (fumaric acid)

The two isomers contain the same functional groups, but the different spatial arrangements of the groups give rise to differences in chemical and physical properties. The most important chemical difference is the relative ease of anhydride formation : the *cis*-isomer readily forms a cyclic anhydride on heating to 170°C.

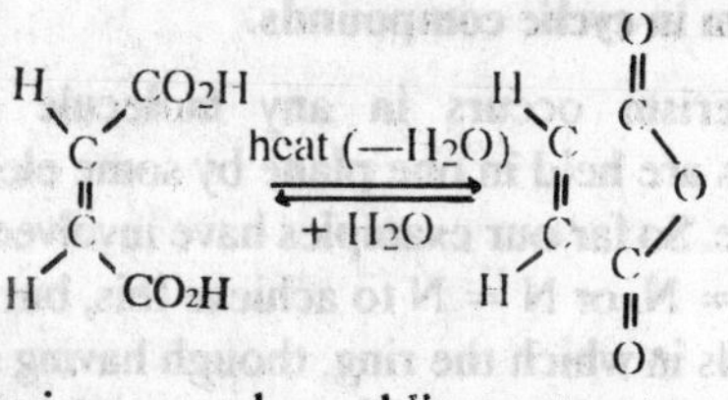

The reaction is reversed on adding water. *trans*- Butenedioic acid remains unchanged these conditions, although on strong heating it sublimes and some rearrangement occurs to yield a certain amount of the *cis*-anhydride.

Reactions such as the formation of the cyclic anhydride, and its subsequent hydrolysis with cold water to yield the *cis*- acid, provide the principal chemical method of distinguishing between geometrical isomers of this type. Physical methods include X-ray crystallography and spectroscopic techniques. The principal differences in the physical properties of *cis*- and *trans*-butenedioic acids are summarized in Table 1. The *cis*-isomer is a considerably stronger acid than the *trans*-isomer for the first ionization owing to stabilization of the resulting anion by intramolecular hydrogen bonding. This order is reversed for the second ionization, for which *trans*-butenedioic acid is approximately 15 times stronger than the *cis*-isomer. The lower acid strength of *cis*-butenedioic acid for this second ionization is partly a result of the stability of the monoanion, but it also reflects the difficulty of removing the second proton from a carboxyl group which is close to the site of the anionic charge.

Table 1. Geometrical isomerism : differences in physical properties of *cis*- and *trans*-butenedioic acids

Property	*cis-Butenedioic acid*	*trans-Butenedioic acid*
m.p. (°C)	130-131	286 (dec.)
density (g cm^{-3})	1.590	1.635
solubility in water (g 100 cm^{-3})	79	0.7
pK_a	1.92	3.02
pK_a''	6.23	4.38
K_a'/K_a''	20.420	23

Cis-trans-isomerism in cyclic compounds.

Cis-trans-isomerism occurs in any molecule in which four appropriate groups are held in one plane by some element of rigidity within the molecule. So far our examples have involved a double bond such as C = C, C = N, or N = N to achieve this, but it also happens in cyclic compounds in which the ring, though having some flexibility, nevertheless restricts the freedom of movement of groups attached to it. In the following discussion the cyclohexane system is treated as planar. This is not so, but its non-planarity does not weaken the argument and it is simpler at this stage to accept this slight deception.

Cyclohexane-1, 4-dicarboxylic acid can be written as structure 1 with the understanding that all the carbon atoms carry as many hydrogen atoms as is required for them to have the usual quadrivalency. However, this compound exists in two forms (mp 162° and 300°) of which only the lower melting isomer readily forms an anhydride. When hydrated the anhydride furnishes the original low melting dibasic acid. This is another example of *cis-trans*-isomerism. The cyclohexane ring can be considered as planar and the four groups ($H.CO_2H.H.CO_2H$) attached to C(1) and C(4) lie in one plane. The two possible structures are shown (2 and 3).

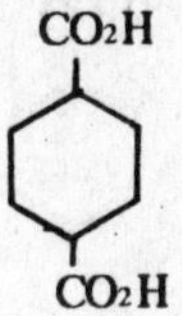

(1)

(2)

(3)

The possibilities of stereoisomerism in compounds with more than one C = C group ?

The possibilities of stereoisomerism increase when the molecule contains more than one C = C (or C = N, etc.) group. Each double bond can lead to an independent set of isomers except that in certain symmetrical molecules some isomers may be identical. Thus the unsymmetrical diene octa-2, 5-diene exists in four stereoisomeric forms and the symmetrical hepta-2, 5-diene in only three.

The structural requirements for *cis-trans*-isomerism

When two atoms are linked by a single bond the barrier to free rotation about this bond is quite small. It is, nevertheless, real and important. On the other hand, when two atoms are multiply bonded then the barrier to free rotation is very much higher and where such compounds exists in two forms the passage from one to the other requires so much energy that each isomer has a stable and independent existence. In short, free rotation about a carbon to carbon double bond cannot occur without disruption of the bond.

In an alkene the two olefinic carbon atoms are sp^2 hybridised and the three sp^2-bonding orbitals lie on the plane. The additional bond formed by overlap of the p_z orbitals greatly restricts the free rotation about the bonds uniting the two olefinic carbon atoms (Fig. 5). The two olefinic carbon atoms and the four attached atoms lie in one plane and, depending on the nature of the attached groups, the alkene may exist in more than one form. This type of stereoisomerism is designated *cis-trans*-isomerism. If the alkene is designated as $abC = Ccd$ then it will exist in two forms so long and $a \neq b$ and $c \neq d$. It does not matter whether *a* and/or *b* are the same as *c*.

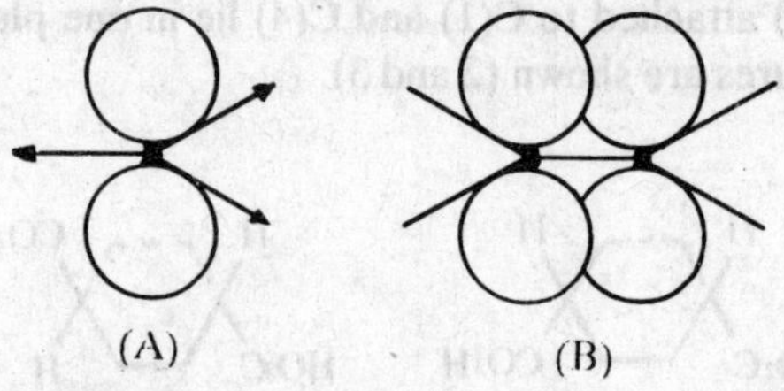

Fig. 5. (A) sp^2-Hybridised cabon atom showing p_z orbital as spheres and three sp^2 hybridised orbitals as lines in one plane and at 120°.
(B) Free rotation in an alkene is restricted because there must be maximum overlap of the two atomic p_z orbitals for maximum stability of the molecular π-orbitals.

It follows, if we are distinguish the two "sides" of the double bond, that groups *a* and *b* must differ from each other and groups *c* and *d* must also differ from each other.

Symbols (E) and (Z)

Difficulty in assigning the terms *cis* and *trans* arises when there is no pair of identical or similar groups. This is now solved by an alternative description which can always be applied. The new terms *(E)* and *(Z)* replace the older terms though it seems likely that *cis* and *trans* will continue to be used to describe alkene isomers for some time yet.

Under the new system the two ligands attached to each olefinic carbon atoms are put in order of precedence by application of the Sequence Rules and the symbols *(E)* and *(Z)* are then used to described those isomers in which the ligands of higher precedence are on the opposite *(E* from the German *entgegen* = *across)* or *the same side* (*Z* from the German *zusammen* = *together*). In many cases, *but by no means all*, the old term *cis* corresponds to *(Z)* and the term *trans* corresponds to *(E)*. This correspondence should never be used without checking. In order to make sure you understand and can use this new description try the following questions.

Is *cis-trans*-isomerism confined to C = C compounds ?

So far we have confined our study of *cis-trans*-isomerism to alkenes but one or both carbon atoms can be replaced by another doubly bonded atom. Most commonly this is nitrogen in which one sp^2 orbital contains only a lone pair. For example, oximes and azo compounds can exist in stereoisomeric forms :

isomeric aldoximes

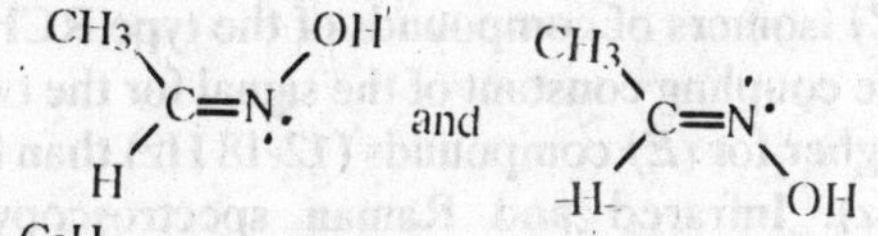

isomeric ketoximes

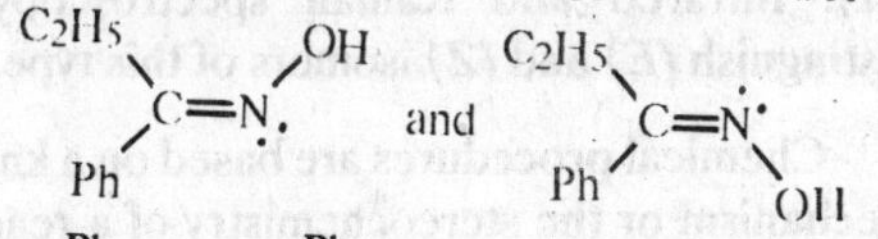

isomeric isomeric azo compounds

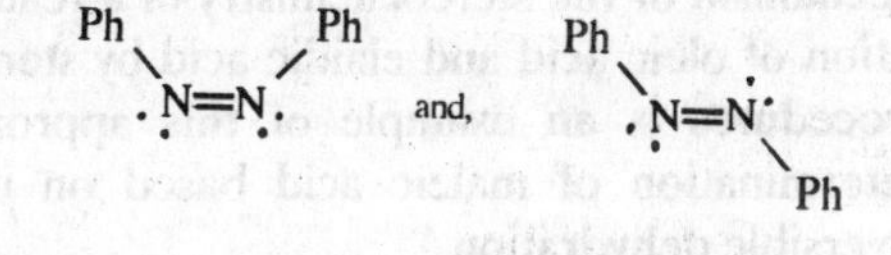

In the past special prefixes have been used to designate these isomers but this is now satisfactorily achieved by the use of *(E)* and *(Z)*.

Assignment of Structures to Geometrical Isomerism

Now we consider briefly the general methods for deciding geometrical confirugration. Most prcedures depend on observations that have been made with a few compounds of known structure and then generalised. This approach must be used carefully and it may not be wise to rely on a single piece of evidence : it is better to seek supporting evidence. The procedures depend on the physical or chemical properties of the two (or more) isomeric forms (these are discussed in greater detail, Brewster, 1972). In general, *cis*- and *trans*-isomers have different physical properties and similar but not identical chemical properties. Chemically the isomers *may* react at different rates on account of steric factors and they *may* give products with different stereochemistry.

Using known compounds many physical properties have been correlated with the *cis*- and *trans*-configuration. Sometimes these correlations are restricted to a group of similar compounds, sometimes they apply more widely. Always they should be used with care. For example, *trans*-compounds often have higher melting points than their *cis*-isomers. Spectroscopic procedures – especially infrared and nuclear magnetic resonance – are usually more reliable. Dipole moments are very satisfactory when they can be applied.

NMR spectroscopy may be used to distinguish between *(E)* and *(Z)* isomers of compounds of the type RCH = CHR′ by attention to the coupling constant of the signal for the two olefinic protons. This is higher for *(E)* compounds (12-18 Hz) than for their (Z) isomers (7-11 Hz). Infrared and Raman spectroscopy may also be used to distinguish *(E)* and *(Z)* isomers of this type.

Chemical procedures are based on a knowledge of either reaction mechanism or the stereochemistry of a reaction product. The preparation of oleic acid and elaidic acid by stereospecific hydrogenation procedures is an example of this approach as is the structure-determination of maleic acid based on its readiness to undergo reversible dehydration.

Interconversion of *cis*- and *trans*-isomers

Early in this chapter it was pointed out that *cis-trans*- isomerism is found among alkenes and certain related nitrogen- containing compounds because the double bond provides a considerable barrier to rotation about its axis. The conversion of *cis*- to *trans*-isomers or the reverse process – known as stereomutation – can be effected in two ways : either by a series of chemical reactions by which the alkene is converted to other compounds from which the alkene can be regenerated as the alternative isomer or by a *reversible* procedure converting the double bond compound to a single bond compound in which free rotation is possible before regeneration of the alkene as a mixture of the two isomeric forms.

This reversible process may be effected by vigorous heating, by photolysis, or, usually under milder conditions, by a free radical species such as those resulting from iodine, thiols or other sulphur compounds, or nitrogen tetroxide. The product is eventually an equilibrium mixture of the two isomers. The *trans*-isomer, being the more stable, usually predominates but pure compounds have to be separated from the reaction mixture.

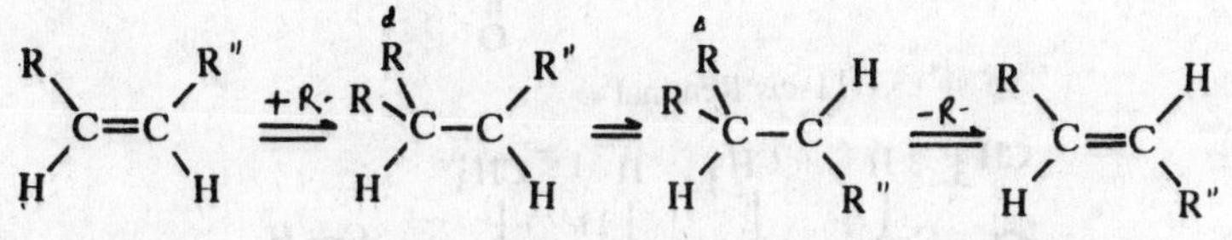

cis - isomer

radical species undergoing inversion through free rotation about the carbon to carbon bond

trans-isomer

Cis-Trans Isomerism in Vision

The retina, the light-detector layer in the back of our eyes, contains colored compounds called *visual pigments*. They are insoluble in water and can be extracted from the retina with aqueous detergents. In the dark these pigments are reddish (their name, rhodopsin, comes from a *Greek* word meaning rose- colored), but the color fades upon exposure to light.

The rhodopsin molecules contains a protein called opsin plus a derivative of vitamin A called 11-*cis*-retinal.

In the dark, 11-*cis*-retinal is stable and fits nicely into the folds of the surrounding opsin. When light hits the rhodopsin, the 11-*cis*-retinal becomes *all-trans*-retinal and no longer fits into the cavity of opsin. The opsin and the *all-trans*-retinal separate. The change in rhodopsin conformation is eventually transmitted to the nerve cells and then the brain. The stereoisomerism of retinal is thus an important part of the vision process.

It is to be noted that only one of the five double bonds is affected in this transformation, but when this one changes from cis to trans, the shape of the entire molecule is changed. An enzyme later catalyzes the change of *all-trans*-retinal back to 11-*cis*-retinal so that it can once again bind to the opsin and wait for the next exposure to light.

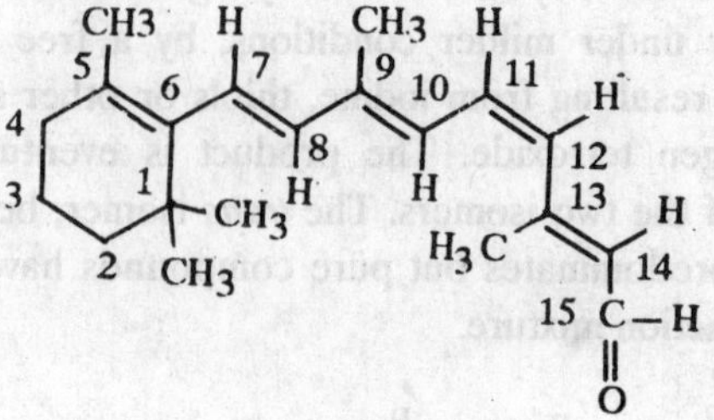

11-*cis* Rentinal

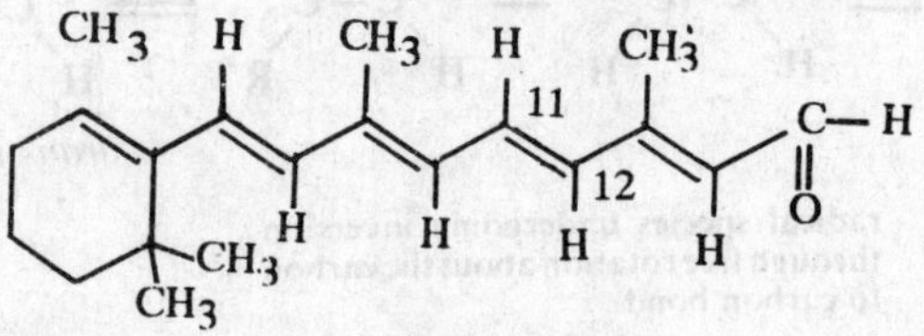

all trams Rentinal

Geometrical Isomerism and Pheromones

Insects emit compounds called **pheromones** to transmit messages to each other. These compounds are remarkably species-specific; that is, each insect is sensitive only to the pheromones of its own species. One of the main uses for these compounds is as sex attractants. Normally, the female emits the pheromone and the male is attracted by its scent. Only a very small quantity is necessary, in some cases as little as 10^{-12} g.

Pheromones are often very simple compounds, though usually with long chains. Two examples are 9-tricosene, the sex attractant of the common housefuly, and bombykol, secreted by the female silkworm moth :

The 9-tricosene secreted by the female housefuly is the cis isomer. The trans isomer does not attract male houseflies and is not secreted by the female. Similarly, the bombykol of the silkworm moth is the trans-cis compound, and none of the three other stereoisomers work. (This is a general rule in nature : in most cases where more than one stereoisomer is possible, nature uses only one).

The discovery of pheromones has given us a possible way to control harmful insects without polluting the environment. Because pheromones have simple structures, they are easily synthesized in

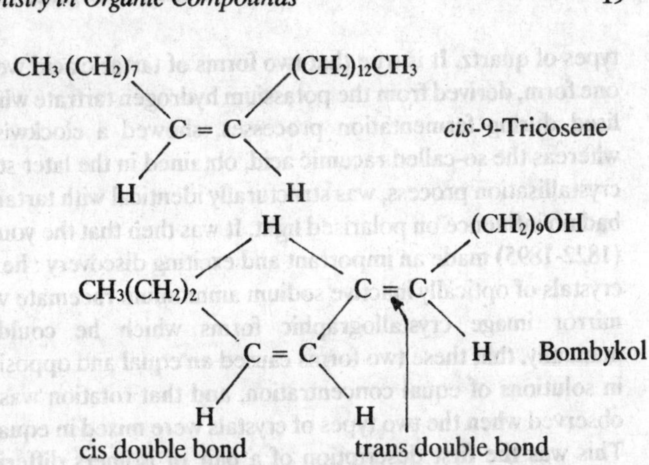

laboratories. A small quantity of a synthetic pheromone can be used to lure male insects into a trap, where they can be killed. Without males, the females cannot propagate. Some of this has already been done – for example, with Japanese beetles – and it is likely that its use will increase.

Optical Isomerism or Enantisomerism

Historical Introduction

The discovery of the polarisation of light by Malus in 1808 was followed by the realisation by Arago (1811) and by Biot (1813) that certain materials deviated the planc of polarisation. Many of these substances were crystalline inorganic compounds such as quartz and sodium chlorate which displayed this property only in the crystalline form. Others, however, were organic compounds, such as oil of turpentine or cane-sugar or camphor which were active in the solid, liquid (whether molten or in solution), and gaseous state. Such materials were described as optically active and it was noted that whilst some rotated plane polarised light in a clockwise fashion other shifted it in an anticlockwise manner.

The crystallographer Haüy had shown that quartz crystals were of two kinds which also differed in their effect on polarised light. One type of crystal produced a clockwise deviation and the other type of crystal an anticlockwise deviation. Up to 1848, however, no organic compound was known to exist in two forms corresponding to the two types of quartz. It is true that two forms of tartaric acid were known : one form, derived from the potassium hydrogen tartrate which crystallised during fermentation processes, showed a clockwise rotation whereas the so-called racemic acid, obtained in the later stages of the crystallisation process, was structurally identical with tartaric acid but had no influence on polarised light. It was then that the young Pasteur (1822-1895) made an important and exciting discovery : he found that crystals of optically inactive sodium ammonium racemate were of two mirror image crystallographic forms which he could separate manually, that these two forms caused an equal and opposite rotation in solutions of equal concentration, and that rotation was no longer observed when the two types of crystals were mixed in equal amounts. This was the first description of a pair of isomers differing only in ability to rotate plane polarised light in opposite directions.

Many refused to believe Pasteur's results and M. Biot, a veteran physicist and the greatest living authority on polarised light, was delegated by the French Academy of Sciences to examine Pasteur's statements. Fortunately we have Pasteur's record of what happend (Read and Gunstone 1958) :

He [M. Biot] sent for me to repeat before his eyes the several experiments. He gave me racemic acid which he had himself previously examined and found to be quite inactive to polarised light. I prepared from it in his presence the sodium ammonium double salt, for which he also desired himself to provide the soda and ammonia. The liquid was set aside for slow evaporation in one of the rooms of his own laboratory, and when 30 to 40 g of crystals had separated he again summoned me to the College de France, so that I might collect the dextro- and laevo-rotatory crystals before his eyes, and separate them according to their crystallographic character, asking me to repeat the statement that the crystals which I would place on his right hand would cause deviation to the right, and the others to the left. This done, he said that he himself would do the rest. He prepared the carefully weighed solutions, and he again called me into his laboratory. He first put the more interesting solution, which was to cause rotation to the left, into the apparatus. Without making a reading, but already at the first sight of the colour-tints presented by the two halves of the field in the Soleil saccharimeter he recognised that there was a strong laevo-rotation. Then the illustrious old man, who was visibly moved, seized me by the hand, and exclaimed :

"My dear child, I have so loved the sciences throughout my life that this makes my heart leap with joy !"

Definition

Some substances are able to rotate the plane of polarized light. This property is called *optical activity*.

Optical activity arises when a molecule cannot be superimposed on its mirror image (Fig. 1 (a)). Such a molecule is described as *chiral*, whereas a molecule which can be superimposed in its mirror image is described as *achiral* (Fig. 1 (b)). Only chiral molecules exhibits optical activity. (It will be useful if you make models of the compounds in Fig. 1 to demonstrate that the two strutures in (a) are different whereas those in (b) are identical).

(a)

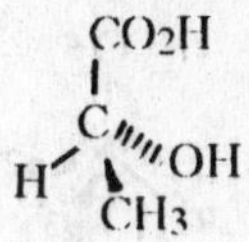

(+)-Lactic acid
((+)-2-hydroxypropanoic acid)

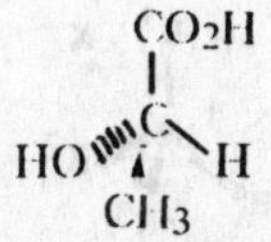

(−)-Lactic acid
((−)-2-hydroxypropanoic acid)

(b)

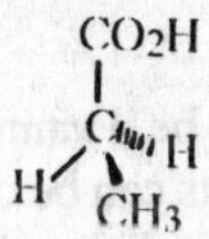

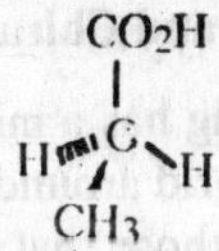

Propanoic acid (two structures identical

Fig. 6. Examples of chiral and achiral molecules

A chiral molecule and its mirror image are different compounds and are therefore isomers. They are called *optical isomers* because they differ in their behaviour towards polarized light. For reasons which will be explained later they are designated (+) and (−) isomers.

Optical isomers such as (+) and (−)-lactic acid which are mirror images of one another are called *enantiomers*. In most respects enantiomers have identical physical and chemical properties. For example, (+) and (−)-lactic acid both have the same melting point and undergo the same chemical reactions with most reagents.

Enantiomers can only be distinguished by their effect upon polarized light and their behaviour towards other chiral species.

The central carbon atom in lactic acid has four different substituents attached to it and is called an *asymmetric carbon atom*. Propanoic acid, which is chiral, does not contain an assymetric carbon atom. The presence of an asymmetric carbon atom in a molecule is a common cause of optical isomerism.

Other atoms besides carbon atoms can also be asymmetric. For example, ammonium salts and amine oxides are tetrahedral in shape, and optical isomers exist when four different substituents are attached to the central nitrogen atom (Fig. 7, R^1, R^2, R^3, and R^4 are alkyl or aryl groups). Compounds containing asymmetric phosphorus and sulphur atoms are also known.

Fig. 7. Compounds containing asymmetric nitrogen atoms

Enantiomers and Chirality

Everything has a mirror image (except may be a vampire). If we look at the world around us, we see that all objects can be divided into two classes : those that can be superimposed on their mirror image and those that cannot. An object that cannot be superimposed on its mirror image is called chiral [pronounced ky' ral]. Chiral objects are nonsuperimposable on their mirror images.

A good example of a chiral object is your left hand. If you hold it up to a mirror, you seen in the mirror an image of your right hand (Fig. 8), and you cannot superimpose the two. Some objects that are not chiral are golf balls, spoons, baseball, bats, and wine classes, all of which are superimpossible on their mirror images (Fig. 9).

The same is true for molecules : Some molecules are chiral, others are not. What kind of molecules are chiral ? Fig. 9, shows a molecule that contains a carbon atom connected to four different groups. (Remember that carbon points its four bonds to the corners to a tetrahedron). The molecule on the left in Fig. 9 is the mirror image of the molecule on the right, just as a left hand is the mirror image of a right hand, but the two molecules cannot be superimposed, any more than we can superimpose a left hand on a right (try putting a left-hand glove on your right hand). A carbon atom connected to four different groups is called a chiral carbon or an asymmetric carbon.

Because the presence of a single chiral carbon atom always makes a molecule chiral, it is important that you be able to recognise one when you see it. It is really not difficult. If a carbon atom is connected to four different groups, *no matter how slight the differences*, it is a chiral carbon. If any two groups are identical, it is not a chiral carbon.

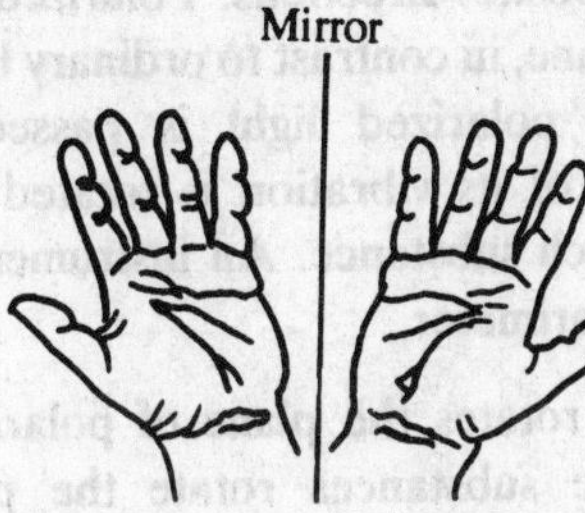

Fig. 8. Left and right hands are nonsuperimposable mirror images and thus are chiral objects.

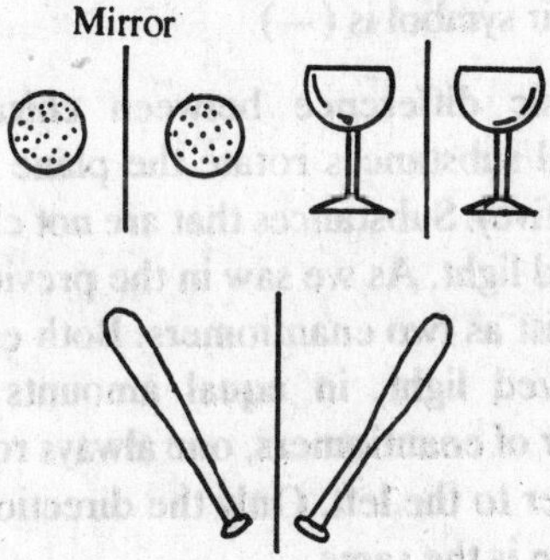

Fig. 9. Some objects that are superimposable on their mirror image. Such objects are not chiral.

Some chiral molecules lack a chiral carbon, but the exceptions are not important enough for us to consider in this book. Also, some molecules that contain two or more chiral carbons are not chiral molecules, but these too are not very important, and we can assume, as a good working rule, that any molecule with one or more chiral carbons is a chiral molecule and any molecule without a chiral carbon is not chiral.

In all cases of isomerism that we have previously seen, both structural isomerism and stereoisomerism, the isomers differ in almost all properties, but this is not the case for enantiomers. All physical and chemical properties of enantiomers are identical, except two.

1. The two members of a pair of enantiomers rotate the plane of polarized light in opposite directions. Polarized light is light that vibrates in only one plane, in contrast to ordinary light, which vibrates in all planes. When polarized light is passed through certain substances, the plane of its vibration is rotated through a certain angle, different for each substance. An instrument that detects this rotation is called a polarimeter.

A substance that rotates the plane of polarized light is called optically active. Some substances rotate the plane to the right (clockwise). We call these dextrorotatory and use the symbol (+). Others rotate the plane to the left (counterclockwise). These are levorotatory, and their symbol is (−).

There is a basic difference between chiral and nonchiral substances. All chiral substances rotate the plane of polarized light (they are optically active). Substances that are not chiral do not rotate the plane of polarized light. As we saw in the previous section, chiral substances always exist as two enantiomers. Both enantiomers rotate the plane of polarized light, in equal amounts but in opposite directions. In any pair of enantiomers, one always rotates the plane to the right and the other to the left. Only the direction is different; the amount of the rotation is the same.

2. Enantiomers undergo chemical reactions, just as any other organic molecule would that has the same functional groups. When they react with another molecule *that is chiral*, the reaction rates are not the same for the two enantiomers. For example, (+)-1-phenyl-1-ethanol reacts with (+)-2-methylbutanoic acid to form a carboxylic ester :

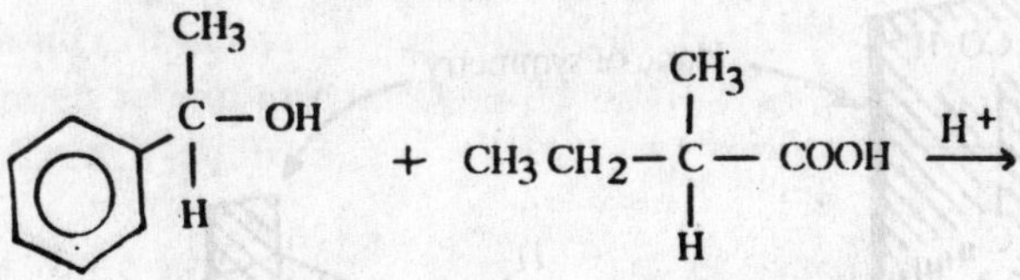

(+)-1-Phenyl-1- ethonal, (+) -2- methybutanoic acid

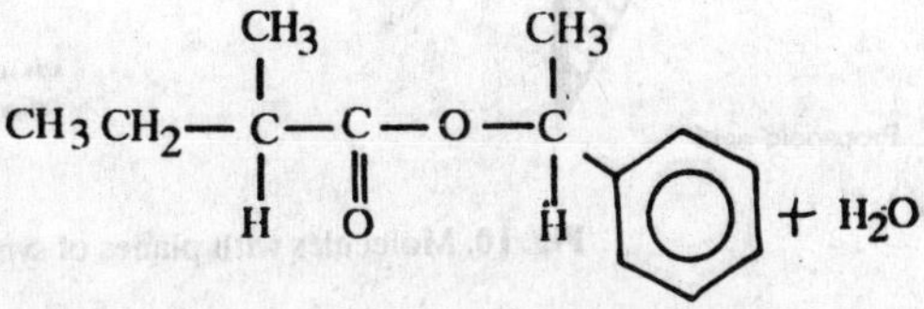

The reaction between (+)-1-phenyl-1-ethanol (–)-2-methylbutanoic acid takes place at a different rate. This is a very important difference in property, with major biological consequences.

In all properties except these two, enantiomers are identical. They have identical melting points, boiling points, densities, rates of reactions with molecules that are not chiral, and all other properties.

Although each enantiomer of a pair rotates the plane of polarized light, a mixture of equal amounts of the two does not because the rotation cancel. Such a mixture is called a racemic mixture.

Elements of Symmetry

The presence or absence of certain elements of symmetry in a molecule can serve as a useful guide to whether a molecule is chiral or achiral. As such it presents an alternative approach to drawing mirror image structures and deciding whether they are superimposable on one another.

The molecule of (+)-lactic acid does not posses a plane or centre of symmetry. It is for this reason that it cannot be superimposed on its mirror image and is chiral. Molecules which do not possess a plane or centre of symmetry are chiral. Molecules which do possess a plane or centre of symmetry are achiral. For example, propanoic acid has a plane of symmetry, as does *cis*-cyclopropane-1,2-dicarboxylic acid (Fig. 10). These are both chiral molecules.

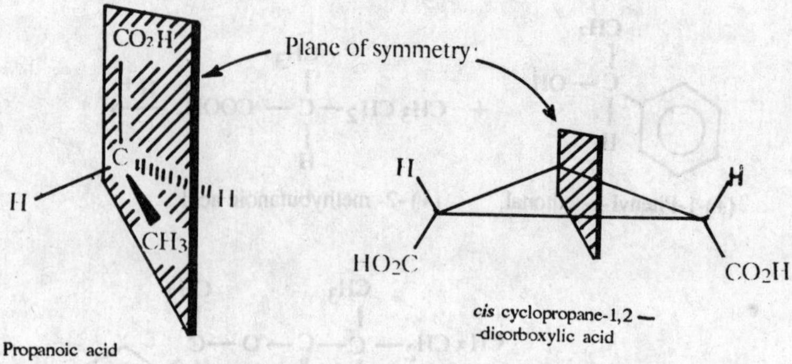

Fig. 10. Molecules with planes of symmetry

The isomer of 2,4-dimethylcyclobutane-1,3-dicarboxylic acid shown in Fig. 11 has no plane of symmetry but it does possess a centre of symmetry as indicated. It is therefore achiral.

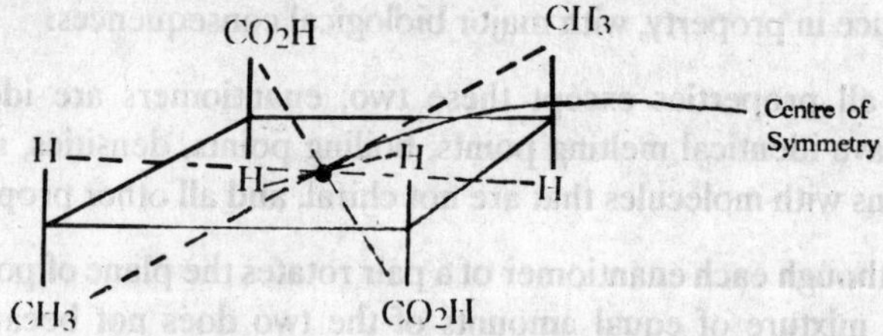

Fig. 11. A molecule with a centre of symmetry.

Optical Isomerism

Sunlight, or light from an ordinary electric lamp, is made up of electromagnetic waves, the vibrations of which take place in any plane

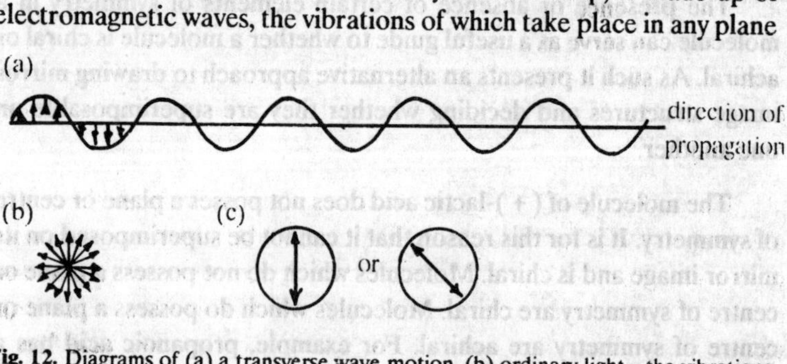

Fig. 12. Diagrams of (a) a transverse wave motion, (b) ordinary light – the vibrations take place in any plane at right angles to the direction of propagation, and (c) plane-polarized light – the vibrations take place in one plane only.

at right angles to the direction of propagation (Fig. 12). If, however, the light passes through a piece of Polaroid or a Nicol prism (a *polarizer*) the waves of the light that emerges are found to be vibrating in *one plane only*, and the light is said to be *plane-polarized*. If this polarized light is now passed through a second piece of Polaroid or Nicol prism (or *analyser*), light will only be transmitted if the plane of polarization is the same as that which the analyser will transmit.

Optical Activity

Ordinary light can be thought of as a series of waves vibrating in random directions perpendicular to the direction of propagation. However if the light is passed through a sheet of Polaroid and Nicol prism (constructed from two crystal of Iceland spar) it becomes plane polarized, *i.e.*, it is reduced to a form in which all the waves vibrate in one plane. A second sheet of Polaroid oriented parallel to the first allows the polarized light to pass through unimpeded (Fig. 13(a)), but if the second sheet is oriented perpendicular to the first then the light beam is completely extinguished (Fig. 13(b)). This effect can be readily observed using two pairs of Polaroid subglasses.

Optically active substances rotate the plane of polarized light. The angle of rotation can be measured by rotating the second sheet at Polaroid until the light beam is extinguished (Fig. 13(c)).

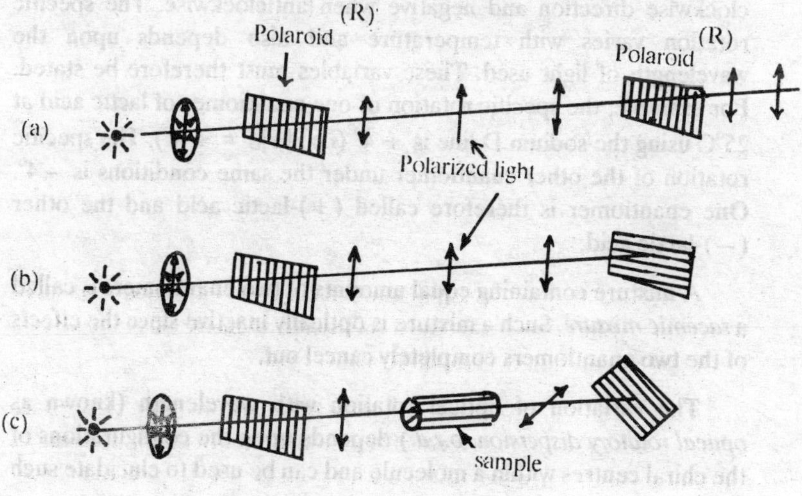

Fig. 13. The rotation of polarized light.

Solutions of certain substances are able to rotate the plane of plane polarized light; such substances are said to be *optically active.* *Optical isomers* are optically active substances which have the same molecular structure but differ in their effect on plane-polarized light. One isomer, the (+) or *dextrorotatory* form, rotates the plane of polarization to the right, while the other rotates it to the left and is known as the (−) or *laevorotatory* form. The two isomers have the same chemical and physical properties, they differ only in their effect on plane-polarized light and in reactions involving other optically active substances.

Optical activity is measured with an instrument known as a *polarimeter* (see below). The *specific rotation* of a substance, $[\alpha]^t_D$ is defined by :

$$[\alpha]^t_D = \frac{100\alpha}{l \times c}$$

where the measurement was carried out at temperature *t* using light from a sodium lamp (the 'D-line' of mean wavelength 589.3 nm), and α is the measured angle of rotation. *l* the thickness of the solution (in decimetres) and *c* the concentration of the solution in grams of solute per 100 cm^3 of solution.

The rotation is defined as positive when it takes place in a clockwise direction and negative when anticlockwise. The specific rotation varies with temperature and also depends upon the wavelength of light used. These variables must therefore be stated. For example, the specific rotation of one enantiomer of lactic acid at 25°C using the sodium D line is + 4° (i.e., $[\alpha]^{25}_D = +4°$). The specific rotation of the other enantiomer under the same conditions is −4°. One enantiomer is therefore called (+)-lactic acid and the other (−)-lactic acid.

A mixture containing equal amounts of two enantiomers is called a *racemic mixture*. Such a mixture is optically inactive since the effects of the two enantiomers completely cancel out.

The variation of optical rotation with wavelength (known as *optical rotatory dispersion, o.r.d.*) depends upon the configurations of the chiral centres within a molecule and can be used to elucidate such configurations.

Specific rotation is a physical property, like density and melting point, and can be used to help identify an unknown optically active compound. Because it also depends on two other variables, the temperature and the wavelength of the polarized light, it is often reported like this (for example, for sucrose) :

$$[\alpha]_D^{20} = +66.4^\circ$$

In this case, the measurement was obtained at 20°C using yellow light from a sodium lamp, for which the letter D is used, following an old custom. This is the type of light used most often.

Of course, if $[\alpha]_D^{20}$ for sucrose is $+66.4^\circ$, then we can be sure that $[\alpha]_D^{20}$ for its enantiomer must be -66.4°.

The polarimeter. The construction of a simple polarimeter – sometimes called a *saccharimeter*, as the instrument is used, especially in the brewing and wine-making industries, for the estimation of sugars – is shown diagramatically in Fig. 14. Light from a monochromatic (single-wavelength) source. *S*, such as a sodium lamp, is polarized by passing it (via a lens) through a Nicol prism known as the *polarizer, P*. The beam of planepolarized light then passes through a glass tube containg the solution or solvent to a second Nicol prim, A (the *analyzer*) which can be rotated.

The tube is first filled with the solvent (which is optically inactive, and is usually water) and the analyser is rotated until the field of view is completely dark. The reading on the analyser scale is noted. If the tube is then filled with a solution of an optically active substance and returned to the instrument, the field of view is no longer dark : the optically active compound has rotated the polarization plane of the beam and the analyser needs to be turned to regain the extinction point. Again the reading on the analyser scale is noted. The angle through which the analyser has to be turned – the difference between the two readings – is the angle of rotation, α.

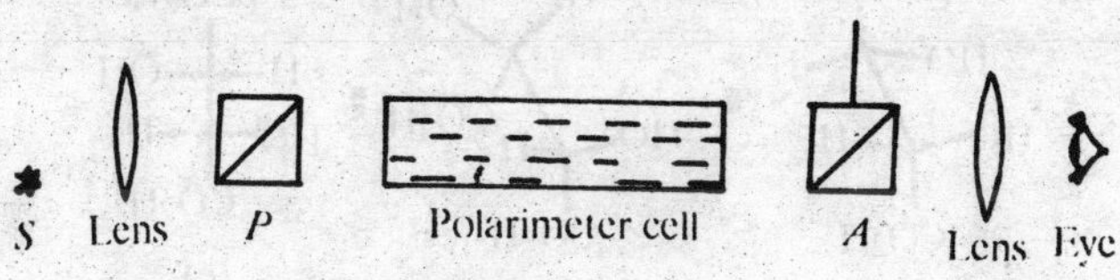

Fig. 14. Diagram of a polarimeter.

Fischer Projection Formulae

Although three-dimensional formulas are very useful for representing the structures of chiral molecules, they are cumbersome to use, especially when the molecule contains several asymmetric centres. To overcome this problem Emil Fischer devised a useful method for representing such structures in two dimensions. Consider first of all a molecule containing only one asymmetric centre (Fig. 15(a)). In order to obtain the *Fischer projection formula* the molecule must first be turned around so that horizontal groups point towards the viewer and vertical groups point away (Fig. 15(b)). Representation of this structure in two dimensions gives the Fischer projection formula (Fig. 15(c)).

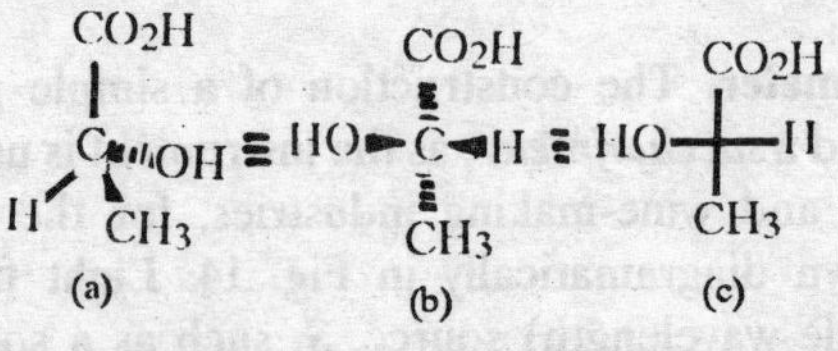

Fig. 15. Construction of a Fischer projection formula

Similar principles apply when drawing Fischer projection formulas of molecules containing more than one asymmetric centre. Thus, in order to draw the Fischer projection formula of (–)-tartaric acid (2,3-dihydroxybutanedioic acid) (Fig. 16(a)) one carbon atom must first be rotated through 180° so that the two carboxyl groups point downwards and the other four groups point upwards (Fig. 16(b)). This structure (which is actually an eclipsed conformation of the molecules – is then viewed from above and its representation in two dimensions gives the Fischer projection formula (Fig. 16(c).

The basic requirements of the Fischer convention are therefore that horizontal lines represent bonds extending above the plane of the

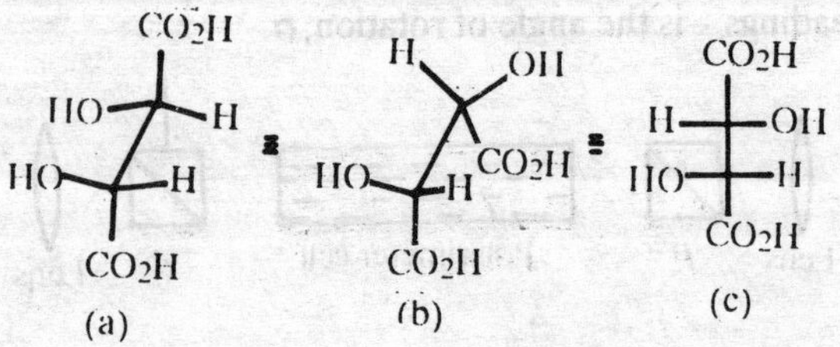

Fig. 16. Fischer projection formula for a molecule with two asymmetric atoms.

paper, and vertical lines represent bonds extending below the plane of the paper. A certain degree of caution must be exercized in handling such formulas. For example, if the projection formula of (+)-lactic acid (Fig. 13(c)) is rotated through 90°, or if any two of the substituents are interchanged, then the formula obtained no longer represents (+)-lactic acid but its enantiomer.

Molecules containing more than one asymmetric centre

When a molecule contains more than one asymmetric centre the number of possible stereoisomers is increased. For example, there are four stereoisomers of 2-hydroxy-3-methylbutanedioic acid (Fig. 17). I and II are mirror images of one another and are therefore enantiomers. Similarly III and IV are enantiomers. However, allthough I and III are optical isomers they are not mirror images of one another. They have the same arrangement of groups (configuration) around one carbon atom but different configurations at the other. They are called *diastereoisomers*. Similarly I and IV are diastereo- isomers, as are II and III, and II and IV. Since diastereoisomers are not mirror images they usually have different physical and chemical properties and can be separated without too much difficulty.

Consider now the case of a molecule which contains two asymmetric centres but has the same four groups around each. In this

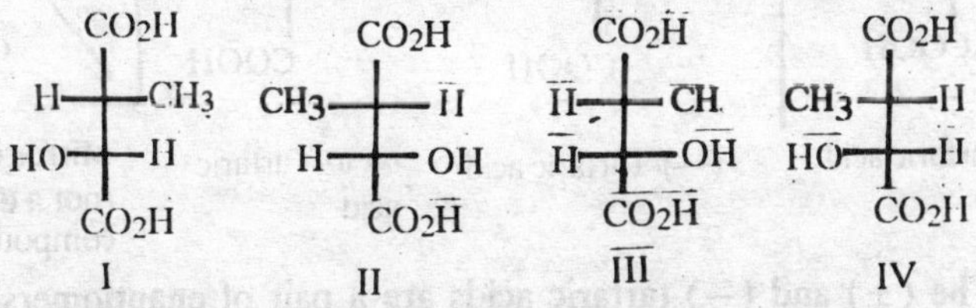

Fig. 17. The stereoisomers of 2-hydroxy-3-methyl butanedioic acid

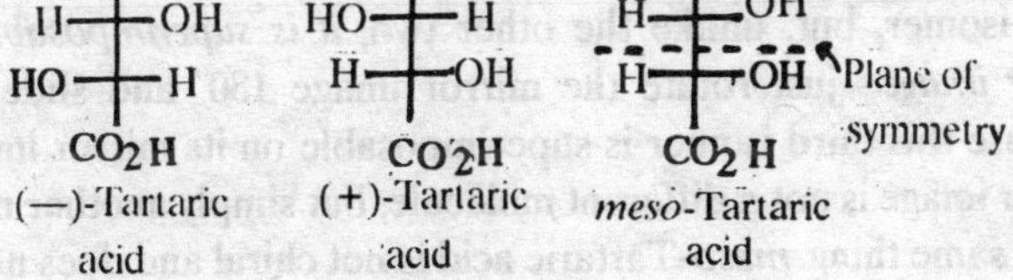

Fig. 18. The stereoisomers of tartaric acid.

case there are only three possible stereoisomers (Fig. 18), because one has a plane of symmetry and is therefore superimposable on its mirror image. This isomer is called the *meso isomer* and is optically inactive.

In a similar manner, it can easily be shown [you may want to try it] that when three chiral carbons are present, there are eight stereoisomers (four pairs of enantiomers), four chiral carbons means 16 stereoisomers, and so forth. The general formula is

$$\text{number of stereoisomers} = 2^n,$$

where *n* is the number of chiral carbon atoms.

The text points out that a compound with two chiral carbons usually has four stereoisomers ($2^2 = 4$). This is indeed the maximum number possible, but in some cases there can be fewer than four. One example is provided by tartaric acid, which has two chiral carbons but only three stereoisomers.

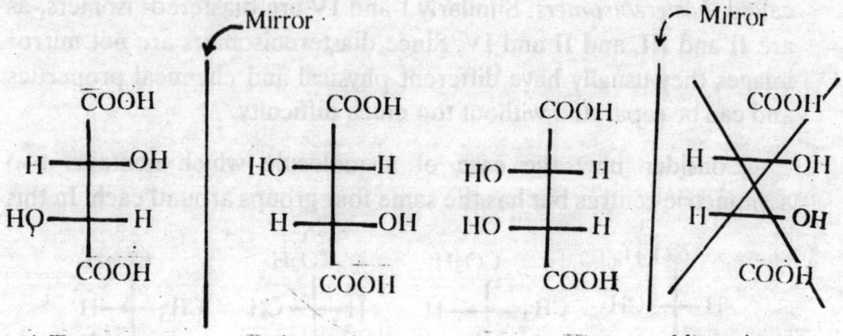

(+)-Tartaric acid (—)-Tartaric acid *meso*-Tartaric acid Mirror image (not a different compound)

The (+) and (−) tartaric acids are a pair of enantiomers and behave exactly the way any pair of enantiomers behave, but look at *meso*-tartaric acid. This compound is obviously not the same as the (+) or (−) isomer – you cannot superimpose it on either one. It is a third isomer, but, unlike the other two, *it is superimposable on its mirror image* – just rotate the mirror image 180° and slide it over. Because this third isomer is superimposable on its mirror image, the mirror image is not a different molecule, but simply another molecule of the same thing. *meso*-Tartaric acid is not chiral and does not rotate the plane of polarized light, even though it has two chiral carbons. It is a diastereomer of the (+) and (−) molecules and so differs from them in all properties.

A meso compound is a compound that possesses chiral carbons (there must be at least two) **but is not itself chiral.**

Optical isomerism with an Asymmetric Atom

In the last section we saw that it was possible for a molecule containing two asymmetric centres to be optically inactive. It is also possible for a molecule which contains no asymmetric centre to be optically active. Thus two enantiomers of penta-2,3-diene exist (Fig. 19). The reason for this is that the molecule as a whole is asymmetric because the two ends are held in perpendicular planes. The molecule must adopt this shape because the central carbon atom is sp^1 hybridized and the two π bonds are therefore orthogonal (i.e. at right angles to each other).

Some *ortho* substituted biphenyl derivatives are also optically active. If the groups in the *ortho* positions are large, rotation about the bond linking the two aromatic nuclei cannot occur. Consequently, when the two substituents attached to each ring are different two discrete isomers exist which cannot be superimposed (Fig. 20).

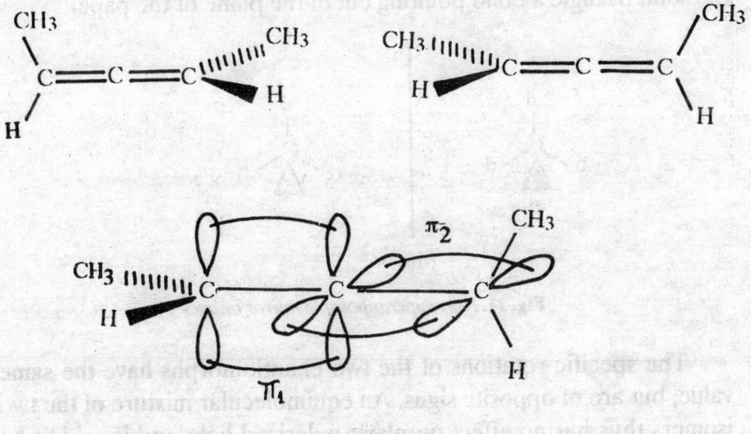

Fig. 19. The two enantiomers of penta-2,3-diene.

which cannot be superimposed (Fig. 20).

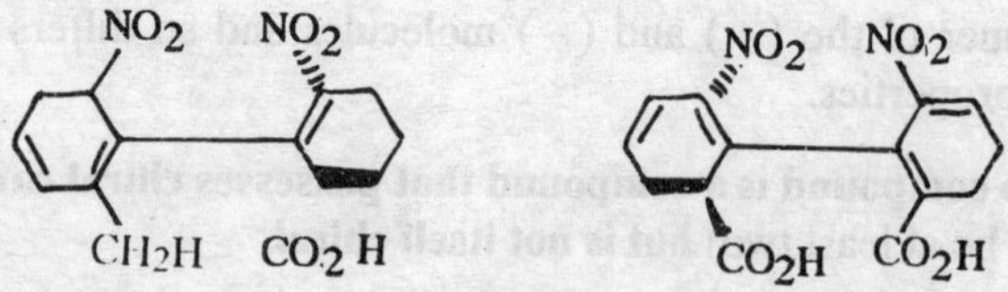

Fig. 20. Isomerism caused by restricted rotation.

Cause of Optical Activity or Molecular Asymmetry

The existence of optical isomers of a substance is a result of the *asymmetry* of its molecules : the criterion for activity is the absence in the molecule of any centre, line or plane of symmetry. The dextrorotatory and laevorotatory isomers have the same structure but are mirror images of each other; because of their asymmetry, the two mirror images are not identical – they cannot be superimposed one on another. The two forms are known as *enantiomorphs* or *enantiomers* and such molecules are said to be *chiral*. The simplest examples are compounds containing one *asymmetric carbon atom*, that is, a carbon atom attached to four different groups, which we may write as *Cabcd*. The bonds are arranged tetrahedrally about the central carbon atom and the configurations of the mirror images are shown in Fig. 21 : the dotted line represents a bond directed into the plane of the page and the solid triangle a bond pointing out of the plane of the paper.

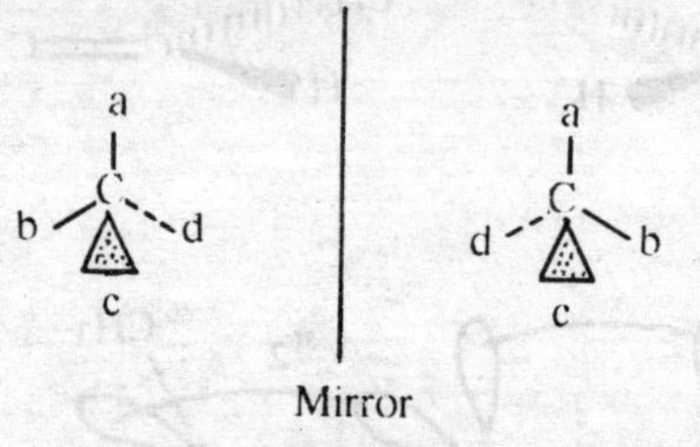

Fig. 21. *Non-superimposable mirror images*

The specific rotations of the two enantiomorphs have the same value, but are of opposite signs. An equimolecular mixture of the two isomers thus has no effect on plane-polarized light and is said to be *optically inactive*. Such a mixture is known as a *racemate* or *racemic mixture* (the DL-form). The letters D and L refer to the configurations of the groups within the molecule.

2-Aminopropanoic acid (alanine). $CH_3CH(NH_2)CO_2H$ and 2-hydroxypropanoic acid (lactic acid). $CH_3CH(OH)CO_2H$, are both examples of naturally occurring optically active substances. L(+)-2-Aminopropanoic acid is an important component of proteins (see Unit 20.9). L(+)-2-Hydroxypropanoic acid is found in muscle, while its D(−)-isomer is a product of the fermentation of sucrose by *Bacillus acidi laevolactiti*, and the racemate occurs in sour milk.

Configuration

(a) **Arbitrary Configuration.** Enantiomorphism – non-superimposable mirror-image asymmetry – is quite common. Our hands have the same structure but different configurations : a left-hand glove will not fit the right hand and vice versa, because the left hand is a non-superimposable mirror image of the right. We can say unequivocally which is the right hand and which is the left. But how can we distinguish between the two isomers of, say, 2-hydroxypropanoic acid ? Which of the two possible configurations is the dextrorotatory and which the laevorotatory isomer ?

The early stereochemists had no physical methods for determining the *actual* positions of groups within a molecule. One substance, (+)-2,3-dihydroxypropanal (dextrorotatory glyceraldehyde), was therefore chosen as an *arbitrary standard* and assigned the configuration shown on the left in Fig. 22. Any substance which could be related by an unambiguous series of reactions to this compound was said to have the D-configuration, irrespective of the direction in which it rotated the plane of polarized light. Similarly, any compound unambiguously related to its mirror image (on the right in Fig. 22) was said to have the L-configuration.

For convenience, these configurations are represented as planar diagrams (such as Fig. 22*(b)* and *(d)*), in which – by convention – the asymmetric carbon atom is at the point where the lines cross, while vertical lines represent groups below the plane of the paper and horizontal lines groups above the plane; where no substitutent group is named in a planar diagram, the presence of a hydrogen atom is understood. The planar diagram of a compound containing an asymmetric carbon atom cannot be superimposed on its mirror image *within the plane of the paper*.

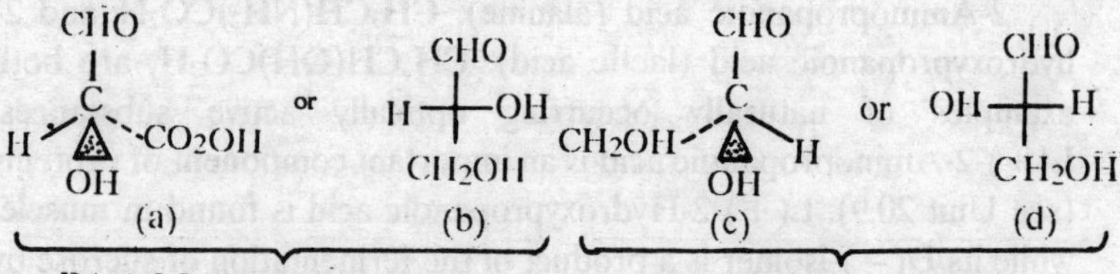

D(+)-2,3-dihydroxypropanal L(−)-2,3-dihydroxypropanal

Fig. 22. *Configurations of 2,3-dihydroxypropanal.*

Compounds containing more than one asymmetric carbon atom, such as the sugars are classified into D- and L-series depending on whether the configuration of the carbon atom at the end of the chain is related to that of D(+)-or L(−)-2,3-dihydroxypropanal :

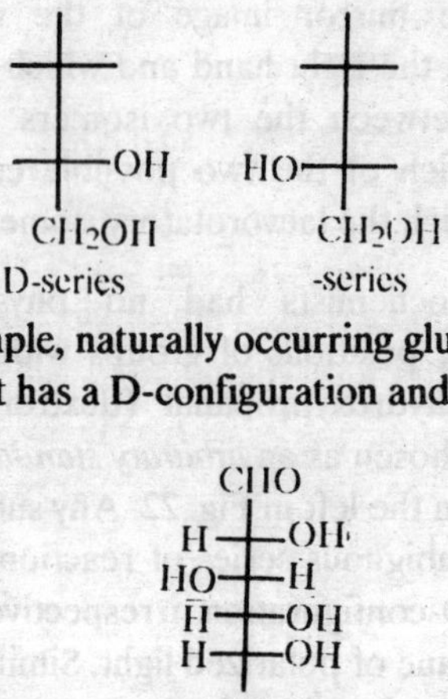

D-series -series

Thus, for example, naturally occurring glucose (Fig. 23) is called D(+)-glucose, as it has a D-configuration and is dextrorotatory.

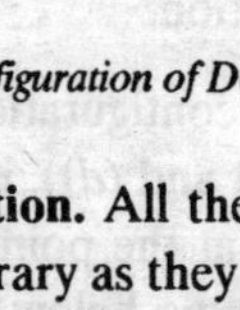

Fig. 23. *Configuration of D(+)-glucose.*

(b) Absolute Configuration. **All the configurations assigned on this system were purely arbitrary as they depended on a guess for the configuration of the standard, D(+)-2,3-dihydroxypropanal. What then if this guess was wrong ? Would all the assigned configurations have to be altered ?**

In 1951, by X-ray crystallographic studies of crystals of sodium rubidium 2,3-dihydroxybutanedioate. Bijvoet was able to check the

actual arrangement of the groups in dextrorotatory 2,3-dihydroxy-butanedioic acid (D(+)-tartaric acid). This substance occurs in grape-juice and had been assigned the configuration shown in Fig. 24 by Emil Fischer about sixty years earlier. Bijvoet's results confirmed that the guess was correct and that the arbitrary configurations based on D(+)-2,3-dihydroxypropanal are actual or *absolute configurations*. The D(+)- and L(−)-dihydroxybutanedioic acids are used as *absolute standards*

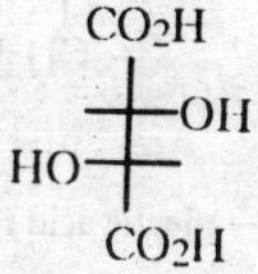

Fig. 24. *D(+)-2,3-Dihydroxybutanedioic acid (D(+)-tartaric acid)*

Nomenclature of Stereoisomers

The configurations of geometrical isomers have been denoted by the terms *cis* and *trans*. It would clearly be useful to have a similar method to denote the configurations of optical isomers. Remember that the prefixes (+) and (−) do not describe the configuration of an isomer but merely indicate the sign of its specific rotation.

For many years the configurations of optical isomers have been denoted by the symbols D and L to indicate their relationship to (+)- or (−)-glyceraldehyde (2,3-dihydroxypropanal) (Fig. 25). Thus compounds having the same configuration as (+)- glyceraldehyde are described as having the D configuration, while those having the same configuration as (−)-glyceraldehyde are described as having the L configuration. The absolute configuration of (+)- glyceraldehyde has been verified by X-ray crystallography.

Glyceraldehyde was originally chosen as the reference compound because it could be converted into many other naturally occurring, optically active compounds. For example, (+)- glyceraldehyde can be converted into (−)-lactic acid in a series of steps which do not affect

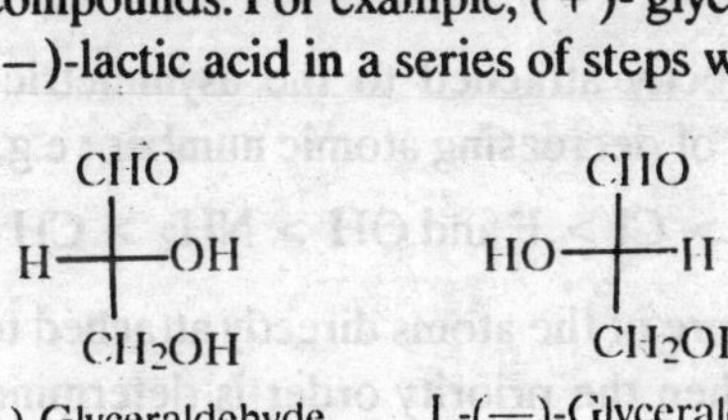

D-(+)-Glyceraldehyde L-(—)-Glyceraldehyde

Fig. 25. The optical isomers of glyceraldehyde

the configuration of the asymmetric centre (Fig. 26). The two molecules therefore have the same configuration (D) although the signs of their specific rotations differ. The D/L system becomes difficult to apply when the groups around the asymmetric centre are not very similar to those in glyceraldehyde.

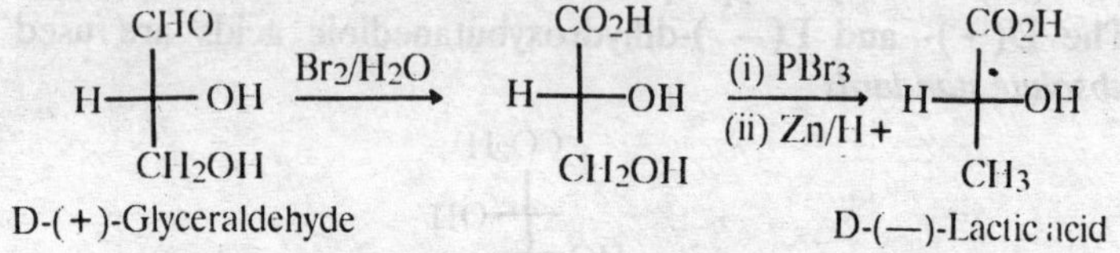

Fig. 26. Synthesis of (−)-lactic acid from (+)-glyceraldehyde

Although the D/L system is often used, particularly in the fields of carbohydrate and amino acid chemistry, a more general method for denoting the configurations of optical isomers has been devised by Cahn, Ingold, and Prelog. The four groups around an asymmetric centre are first placed in an order of priority (a,b,c,d) using a set of rules given below. The molecule is then viewed from the side opposite the group of lowest priority (Fig. 27). If the remaining groups (a,b,c) are arranged clockwise as seen from this viewpoint then the configuration is designated *R* (Latin, rectus = right), whereas if they are anticlockwise it is *S* (Latin, sinister = left).

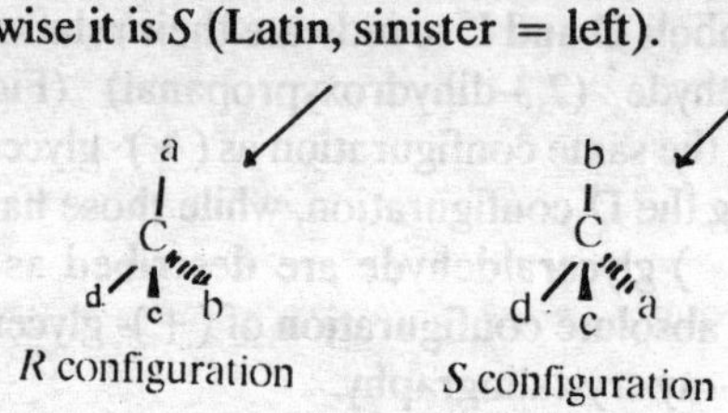

Fig. 27. The *R/S* designation

The priority order (a,b,c,d) is decided as follows :

1. Atoms directly attached to the asymmetric centre are first arranged in order of decreasing atomic number : e.g.

$$I > Br > Cl > F \text{ and } OH > NH_2 > CH_3 > H.$$

2. If two or more of the atoms directly attached to the asymmetric centre are alike then the priority order is determined by the atomic numbers of the second or subsequent atoms : e.g.

$$CH_2OH > CH_2NH_2 > CH_2CH_3 > CH_3 \text{ and}$$
$$CH_2CH_2OH > CH_2CH_2CH_3 > CH_2CH_3.$$

3. If the second (or subsequent) atoms are the same but the number of such atoms is different, then the group with more substituents of higher atomic number takes priority : e.g.

$$CHCl_2 > CH_2Cl$$

4. When the atom attached to the asymmetric centre is multiply bonded then the atom at the other end of the multiple bond is counted twice (for a double bond) or three times (for a triple bond) : e.g.

$$CO_2H > CHO > CH_2OH.$$

On the basis of the above rules the priority order for the four groups attached to the asymmetric centre in lactic acid is OH > CO_2H > CH_3 > H. Consequently (+)-lactic acid has the *S* configuration (Fig. 26). Similarly it can be seen that (+)-glyceraldehyde has the *R* configuration.

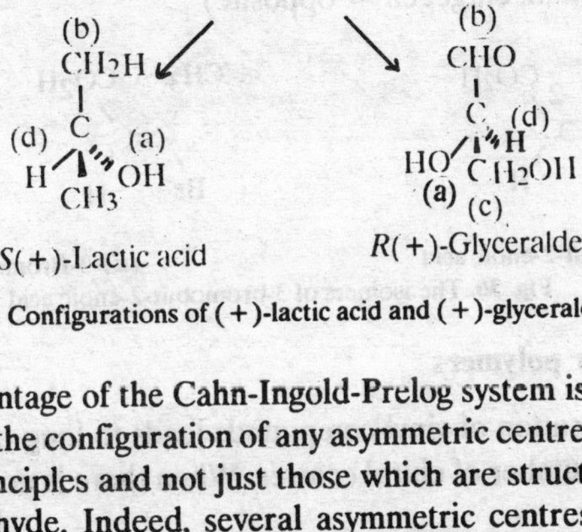

Fig. 28. Configurations of (+)-lactic acid and (+)-glyceraldehyde

The advantage of the Cahn-Ingold-Prelog system is that it allows one to define the configuration of any asymmetric centre in a molecule from first principles and not just those which are structurally related to glyceraldehyde. Indeed, several asymmetric centres in the same molecules can be separately defined by this method. For example the

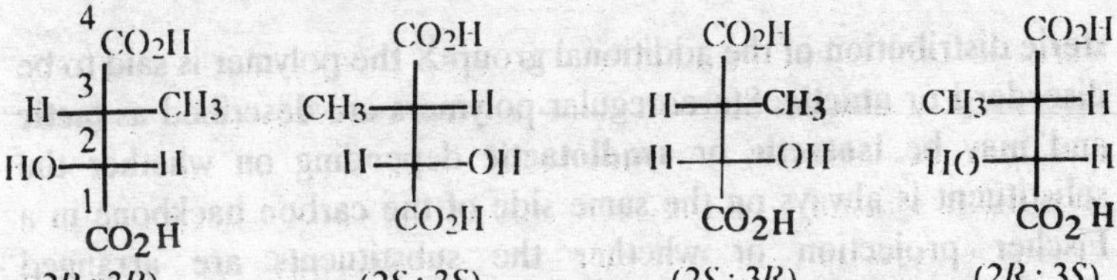

Fig. 29. The isomers of 2-hydroxy-3-methylbutanedioic acid.

configurations of the isomers of 2-hydroxy-3-methylbutanedioic acid (Fig. 29) are completely defined by adding the appropriate prefix, (2*R* : 3*R*), *(2S : 3S)*, *(2R : 3S)* or *(2S : 3R)*, to the systematic name.

In most simple cases the *cis/trans* terminology can be used to define the configuration of geometrical isomers. However in some cases, such as when four different groups are attached to the double-bonded carbon atoms of an alkene, this sytem is difficult to apply. The problem can be overcome by making use of a more general system of nomenclature based on the Cahn-Ingold- Prelog priority rules. One must first assign an order of priority to the groups attached to each carbon atom of the double bond. For example in the isomers of 3-bromobut-2-enoic acid (Fig. 30), the order of priority at carbon atom 3 is Br > CH_3, and at carbon atom 413 CO_2H > H. When the two groups of highest priority are on the same side of the molecule it is called the *Z* isomer (German, zusammen = together) whereas when they are on opposite sides of the molecules it is called the *E* isomer (German, entgegen = opposite).

(Z)-2-Bromobut-2-enoic acid *(E)*-3-Bromobut-2-enoic acid

Fig. 30. The isomers of 3-bromobut-2-enoic acid

Stereoregular polymers

Polymerisation of vinyl compounds leads to long-chain polymers with a large number of chiral centres. When there is no pattern in the

$$CH_2{=}CHX \longrightarrow CH_2CH(X)CH_2CH(X)CH_2CH(X)CH_2CH(X)$$

steric distribution of the additional group X the polymer is said to be disorderd or **atactic.** Stereoregular polymers are described as **tactic** and may be **isotactic** or **syndiotactic** depending on whether the substituent is always on the same side of the carbon backbone in a Fischer projection or whether the substituents are arranged alternately. In reality, the long-chain molecules tend to assume a helical form.

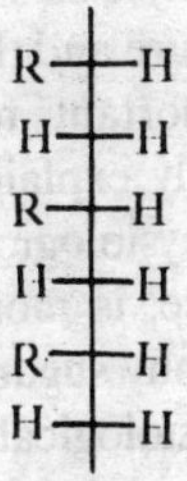

isotactic

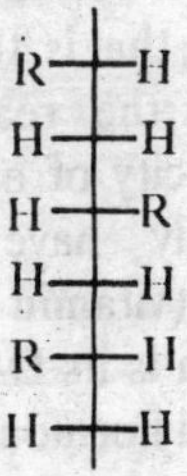

syndiotactic

Organometallic catalysts furnish products with a high degree of stereospecificity and polypropylene, polybutene, and polystyrene, exist as isotactic polymers. Polybutadiene and polypropylene prepared under other conditions are syndiotactic. Stereoregular polymers have exceptionally high crystallinity, density, and melting point.

Properties of Enantiomers

(a) **Physical Properties.** The physical properties (melting point, boiling-point, solubility acid, strength, density and so forth) of any pair of isomers are all identical, except for their effect on plane-polarized light.

(b) **Chemical Properties.** The chemical properties of optical isomers are identical, except in some reactions that involve the participation of other optically active or *stereospecific* substances. Many enzyme-catalysed reactions occur readily with one isomer while its mirror image reacts either very slowly or not at all. For example, lactate dehydrogenase, an enzyme found in muscle, catalyses the dehydrogenation of L(+)-2-hydroxypropanoic acid to 2-oxopropanoic acid :

$$CH_3{}^*CH(OH)CO_2H \xrightarrow[(+2H)]{\text{lactate dehydrogenase}} CH_3COCO_2H$$

L(+)-2-hydroxypropanoic acid (lactic acid) — 2-oxopropanoic acid (pyruvic acid)

where *C represents an asymmetric carbon atom (the *chiral centre*). The same enzyme catalyses the reverse reaction, but only the L(+)-acid is obtained from the optically inactive 2- oxopropanoate.

Reactions of this type indicate that a similar *stereospecificity* exists in the structure of enzyme molecules, that is, that both the nature and the stereochemical configuration of the reactant are important for enzyme action. The stereospecificity of enzymes probably explains why optical isomers frequently have different physiological properties; L(+)- ascorbic acid (vitamin C), for instance, is more effective in preventing scurvy than is its D-isomer, and many sugars, amino-acids, antibiotics, drugs and hormones are only physiologically active in one form.

Can enantiomers be interconverted ?

The short answer to this question would be "no" for there is no simple and general procedure for converting a stereoisomer to its enantiomeric form. Nevertheless this answer is not quite complete and there are two additional points to be made.

It is sometimes possible to devise a series of reactions, all of which occur with a high degree of stereospecificity, so that the overall result of the sequence is to change one enantiomer to the other.

In addition there are several procedures whereby an enantiomer [(+)-A] can be converted to its racemate [(±)- A] and then separated into its two constituents [(+)-A and (−)- A]. These two processes are described as racemisation and resolution respectively.

Racemisation is sometimes achieved by vigorous heating or through the catalytic activity of an acid, a base, or some metallic compound or by conversion to an achiral compound and regeneration in the racemic form. As an example of this last procedure an enantiomeric alcohol can be oxidised to an achiral ketone and then reduced to the alcohol in racemic form.

(+)-RCH(OH)R′ ⟶	RCOR′ ⟶	(±-RCH(OH)R′
chiral molecule enantiomer	achiral molecule	chiral molecule racemic mixture

The alcohol function must, of course, be part of the chiral centre otherwise it would be destroyed during oxidation.

A familiar type of enantiomeric compound which is readily racemised is one in which the chiral atom is attached both to hydrogen and to a carbonyl group. Such compounds are chiral in the keto form but achiral in the enol form. Since regeneration of the keto form is not stereospecific the interconversion is accompanied by racemisation. Both acids and bases catalyse the keto-enol equilibrium and therefore promote racemisation. Enantiomers of this type may not, therefore, be optically stable and may racemise – slowly or quickly – when stored.

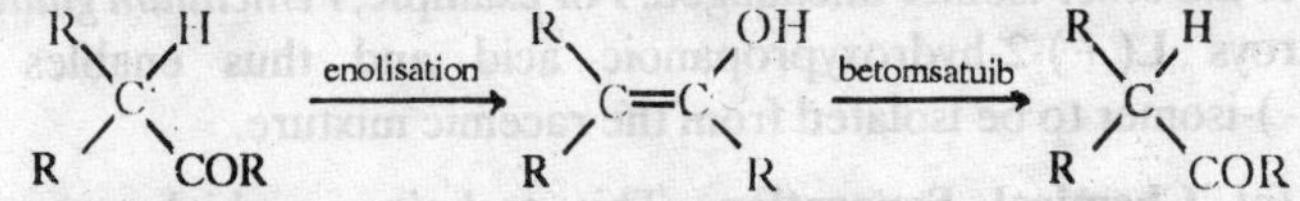

(+)-keto form (chiral) enol form (achiral) (+)-keto form (chiral)

Resolution of Optical Isomers

Some optically active substances may be isolated in the form of a pure enantiomer from natural sources or from the products of an enzyme reaction. But when a compound containing an asymmetric carbon atom is synthesized in the laboratory from optically inactive starting materials, the product is almost invariably a racemic mixture. 2-Hydroxypropanoic acid, for example, is obtained from ethanal as a racemate by the following reaction sequence (since there is an equal chance of attack on either side of the planar carbonyl group) :

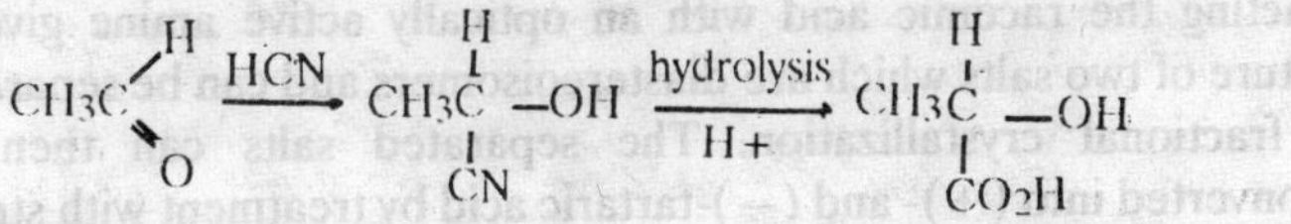

enthanal DL-2-hydroxy-pranonitrile DL-2-hydroxypropanoic acid

Pasteur performed the first *resolution* of an optically active compound by manually separating the crystals of (+)- and (−)-lactic acid under a microscope. However this method can only be used in a limited number of cases and is extremely tedious.

There are three principal methods of separating or *resolving* such a racemic mixture into the optical isomers.

(a) **Mechanical Separation.** This technique, often called *crystal picking*, was first used by Pasteur in 1848 to separate the two forms of sodium ammonium 2,3-dihydroxybutanedioate and is of mainly historical interest. The method depends on the fact that the crystals of

the two enantiomers of this salt are themselves enantiomorphous – an unusual phenomenon – and can be separated with tweezers under a lens or a low-powered microscope.

(b) **Biochemical Separation.** Occasionally it is possible to find a micro-organism which metabolizes one form in a racemic mixture and leaves the other isomer unchanged. For example, *Penicillium glaucum* destroys L(+)-2-hydroxypropanoic acid and thus enables the D(–)-isomer to be isolated from the racemic mixture.

(c) **Chemical Separation.** This technique, which was also developed by Pasteur, is generally the most convenient. It usually depends on salt formation : a racemic acid, for example, is treated with an enantiomer of an optically active base, generally a compound that occurs naturally in optically pure form, such as brucine, quinine, strychnine or morphine :

D-acid + L-acid + 2D-base → (D-acid-D-base) + (L-acid-D-base)
racemic mixture

The product consists of a mixture of isomeric optically active (but not enantiomorphous) salts which, if their solubilities are sufficiently different, may be separated by fractional crystallization. The separate acid isomers are then regenerated by hydrolysis of the salt. Consider for example the resolution of a mixture of (+)-and (–)-tartaric acid. Reacting the racemic acid with an optically active amine gives a mixture of two salts which are diastereoisomers and can be separated by fractional crystallization. The separated salts can then be reconverted into (+)- and (–)-tartaric acid by treatment with strong acid.

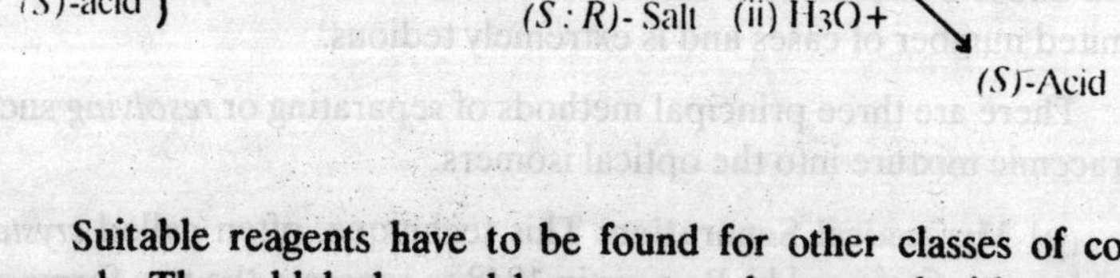

Suitable reagents have to be found for other classes of compounds. Thus aldehydes and ketones may be treated with enantiomeric reagents such as menthylhydrazine or menthylsemicarbazide to give diastereoisomeric products. The isocyanate PhCH(Me)NCO has been used for the resolution of chiral amines and alcohols and the latter have often been separated after conversion to acidic compounds such as the half esters of phthalic acid thus :

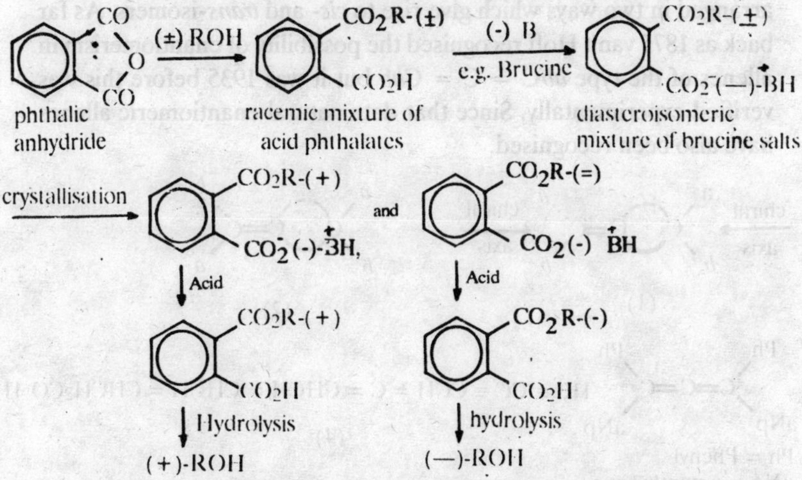

In a recently described analytial procedure for determining the amount of (+)- and (−)-alcohols in a mixture these were esterified with (−)-methyl chloroformate and the diastereoisomeric menthoxycarbonyl esters were separated by GLC. Good separations were effected, though in this case the diastereoisomers were not collected nor were the enantiomers regenerated.

(±)-ROH + (—)-menthylchloroformate → (+)-ROCO$_2$menthyl-(—) and (—)-ROCO$_2$menthyl-(—) (diastereoisomeric menthoxy-carbonyl esters)

Optically Active Compounds with an Axis of Chirality

This chapter has been concerned so far only with compounds which are optically active because they contain one or more centres of chirality. We now direct our attention to compounds having an axis or plane of chirality and no centre of chirality. We shall discuss two types of compounds with an axis of chirality : the allenes and certain related compounds on the one hand and the biphyenyls on the other.

Allenes and Related Compounds

When considering *cis-trans*-isomerism we noted that in the alkene *ab*C = C*ab* the four substituents *abab* lie in one plane and can be arranged in two ways which give rise to *cis*- and *trans*-isomers. As far back as 1875 van't Hoff recognised the possibility of enantiomerism in allenes of the type *ab*C = C = C*ab* but it was 1935 before this was verified experimentally. Since that date natural enantiomeric allenes have also been recognised.

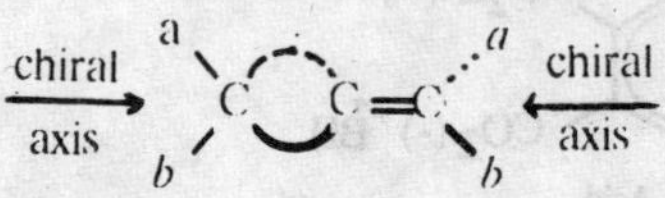

(1)

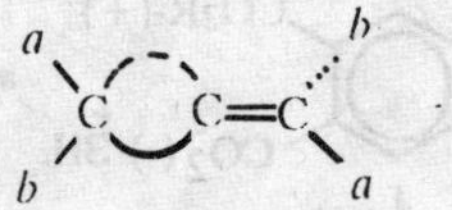

(2)

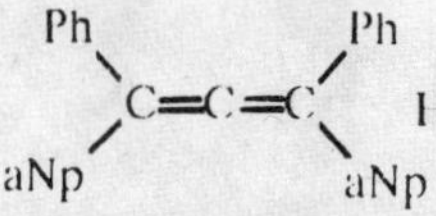

Ph = Phenyl
aNp = a-napthyl

(3)

$$HC \equiv CC \equiv CCH = C = CHCH = CHCH = CHCH_2CO_2H$$

(4)

In an allene the four substituents *abab* attached to the allenic carbon atoms lie in intersecting planes and the two structures 1 and 2 are non-superimposable. The chiral axis in 4 is shown by the two arrows. Typical examples of enantiomeric allenes are the synthetic hydrocarbon 3 and the natural antibiotic mycomycin (4).

Since a ring system can produce the same element of rigidity in a molecule one or both of the double bonds of the allene system can be replaced by rings, and compounds such as (5) and (6) also exist in enantiomeric forms because they contain an axis of chirality.

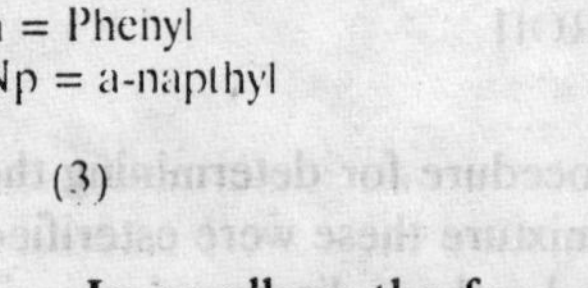

(5)

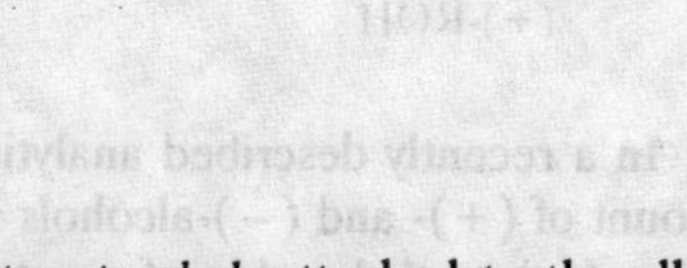

(6)

Biphenyls

In the molecule of biphenyl there exists a possibility of rotation about the single bond linking the two phenyl groups so that biphenyl could conceivably exist in a form (1) in which the two rings are coplanar, in a form (2) in which the two rings are perpendicular to each other, and in any form between these two extremes.

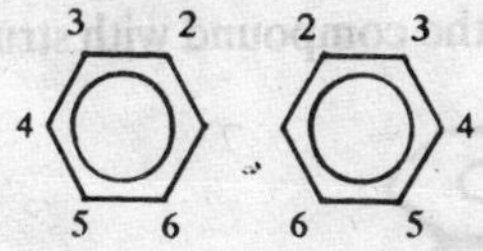

(1)

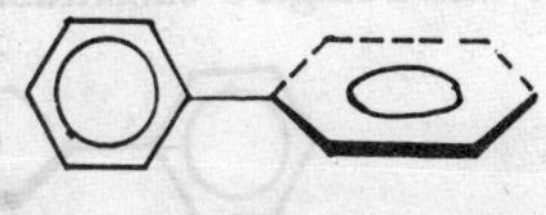

(2)

In the coplanar form the overlap of π orbitals of the two rings allows maximum resonance stabilisation but there is some inhibiting steric interaction of the *o*-hydrogen atoms (2 and 2′, 6 and 6′). In the perpendicular form resonance stabilisation is at a minimum but so also is steric interaction. Electron diffraction data suggest that the two rings are normally inclined at about 40°.

The steric interactions become more significant as the *o*-hydrogen atoms are replaced by bulkier groups and in diphenic acid (3), for example, the angle between the two rings is about 60°.

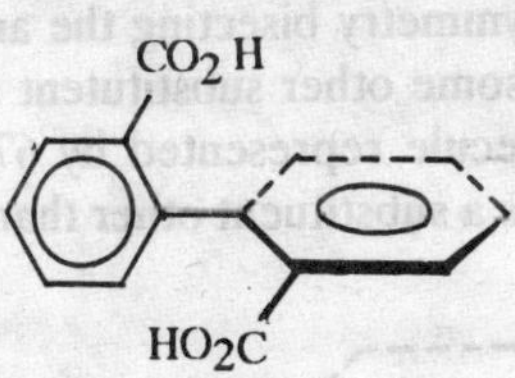

(3)

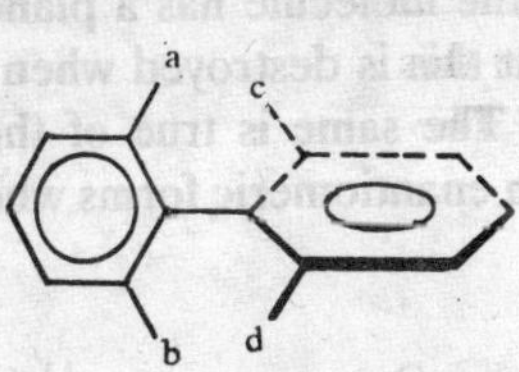

(4)

In a suitably substituted biphenyl compound the non-planar form exists in two enantiomeric structures and since the steric hindrance to rotation depends on the size of these substitutents enantiomeric biphenyls vary in their optical stability.

In summary, there are two conditions for enantiomerism in the biphenyl (4) :

(i) The molecule must not contain a plane of symmetry : this requirement is met either if $a \neq b$ and $c \neq d$ or if $a \neq b$ and $c = d$ with a *m*-substituent in this ring. (ii). There must be a sufficiently large barrier between the two enantiomeric forms to preclude rapid inter-conversion. This is not an all-or-none effect ; the rate of inter-conversion depends on the nature of the *o*-substituents and compounds of varying optical stability are known.

Most enantiomeric biphenyls have three or four substituents in the *o*-positions to provide the necessary optical stability. An unusual example with a single *o*-substituent is the compound with structure 65.

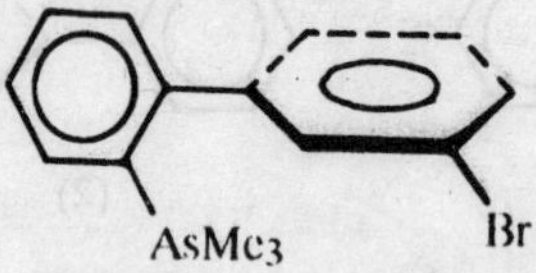

(5)

Optically Active Compounds with a Plane of Chirality

If a molecule has a plane of symmetry which is destroyed by the presence of a suitably placed group then the molecule has a plane of chirality and can exist in enantiomeric forms. In the compound represented by 66, for example, the alicyclic ring is too small to allow the formation of a planar configuration and the aromatic ring therefore takes up an orientation perpendicular to the alicyclic ring. When R = H the molecule has a plane of symmetry bisecting the aromatic ring but this is destroyed when R is some other substitutent such as CO_2H. The same is true of the molecule represented by 67 which exists in enantiomeric forms when R is a substituent other than H.

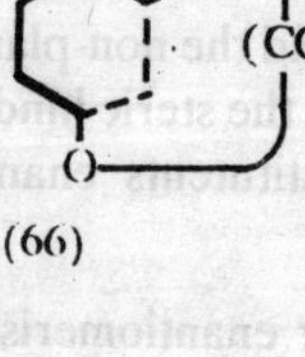

(66)

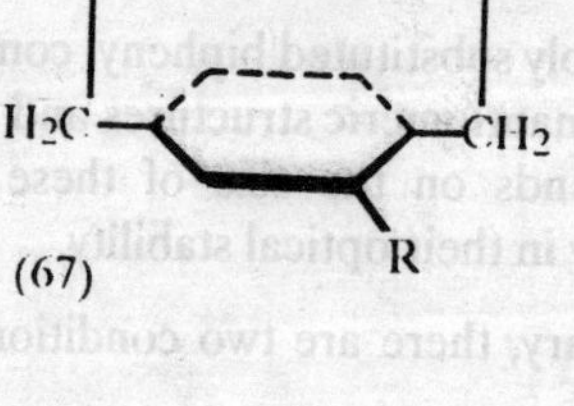

(67)

Helicenes

An unusual example of chirality is provided by the helicenes. As this trivial name suggests, these molecules assume a helical or spiral arrangement and exist in enantiomeric forms depending on whether they are like a right- or left-handed screw. Of added interest is the fact that these compounds have the highest optical rotation yet recorded.

In the simplest helicene made of six *o*-fused benzene rings planarity is prevented by the presence of carbon atoms which, in a planar arrangement, would occupy the same position and the molecule takes up a helical form. Compounds up to 13-helicene are known and have been described as molecular spiral staircases. Enantiomeric helicenes have been obtained by resolution (using an enantiomeric complexing agent), by the selection of enantiomeric crystals, or by homologation of enantiomeric precursors.

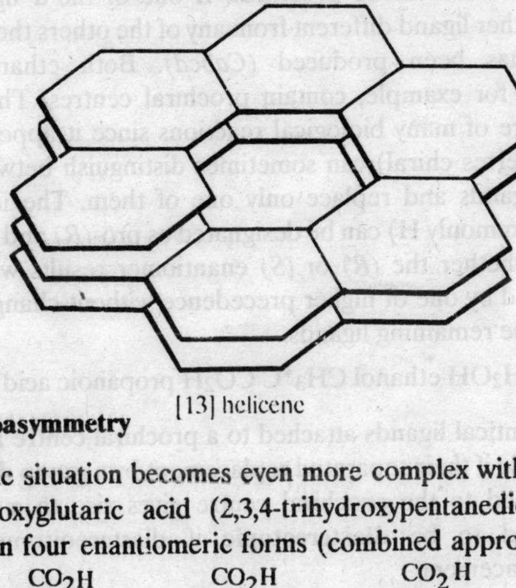

[13] helicene

Pseudoasymmetry

The situation becomes even more complex with a molecule like trihydroxyglutaric acid (2,3,4-trihydroxypentanedioic acid) which exists in four enantiomeric forms (combined appropriately into two

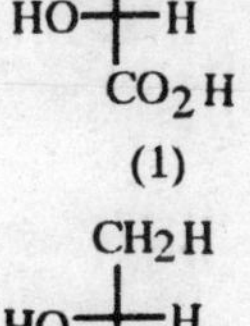
(1)

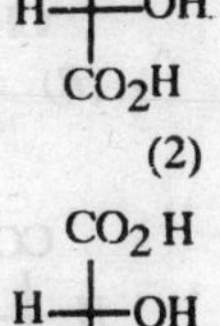
(2)

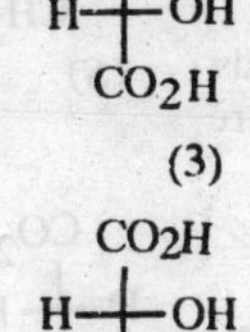
(3)

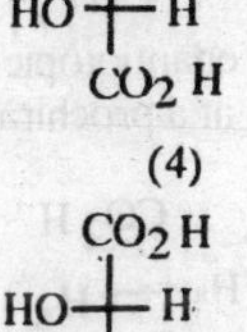
(4)

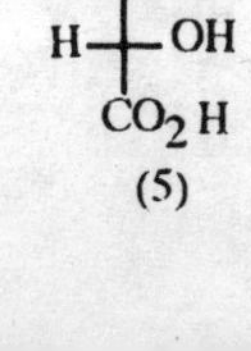
(5)

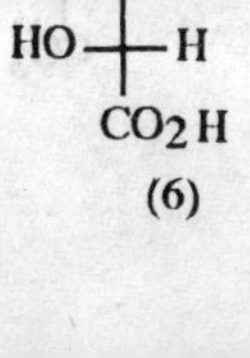
(6)

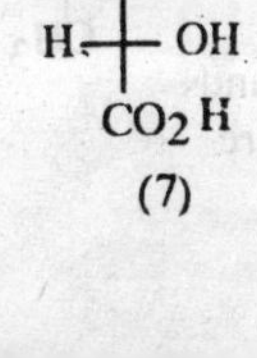
(7)

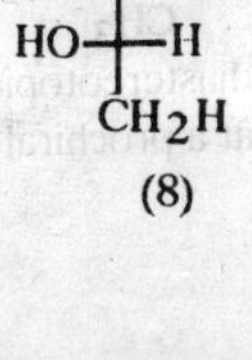
(8)

racemates) and in two different *meso*-isomers. The molecule contains three chiral centres and we can therefore write eight Fischer projections thus :

The central carbon atom in this molecule is linked to two different ligands and to an enantiomeric pair of ligands, *i.e.*, it may be written as $C(+a)(-a)bc$ or as $C(+a)(+a)bc$. Such an atom is said to be pseudoasymmetric when it has the form $C(+a)(-a)bc$.

Prochiral Centres

An atom linked to two identical ligands and to two other different ligands *(Caabc)* is said to be **prochiral**. If one of the *a* ligands is replaced by another ligand different from any of the others then a new chiral centre has been produced *(Cabcd)*. Both ethanol and propanoic acid, for example, contain prochiral centres. This is an important feature of many biological reactions since it appears that enzymes (themselves chiral) can sometimes distinguish between the two identical ligands and replace only one of them. The identical ligands (most commonly H) can be designated as pro-*(R)* and pro-*(S)* depending on whether the *(R)* or *(S)* enantiomer results when the ligand is replaced by one of higher precedence without changing the precedence of the remaining ligands.

CH_3*C H_2OH ethanol CH_3*C CO_2H propanoic acid

The two identical ligands attached to a prochiral centre are said to be **enantiotopic** if their separated replacement by a group different from any attached to the prochiral centre gives rise to a pair of enantiomers and to be **diastereotopic** if diastereoisomers are obtained on replacement.

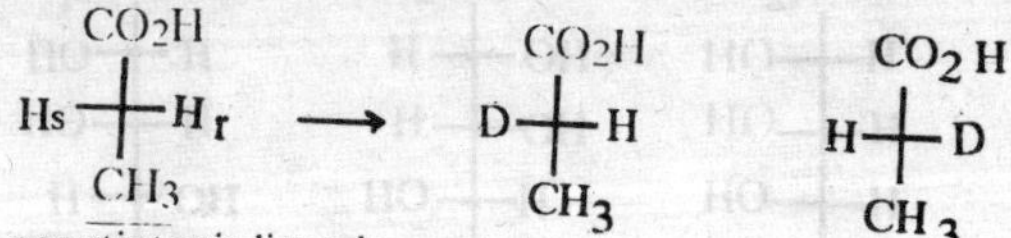

enantiotopic ligands at a prochiral centre (S) (R) (enantioners)

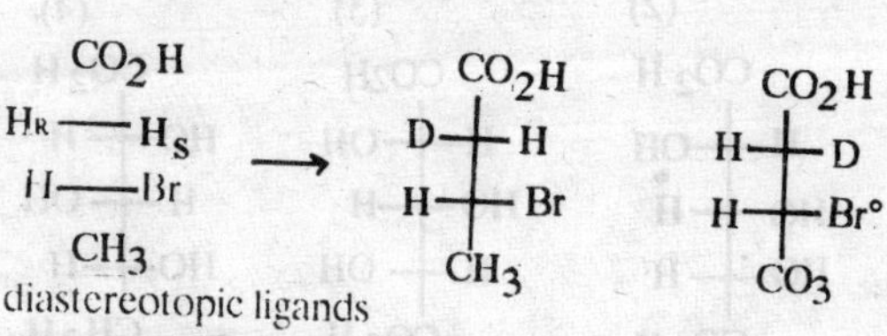

diastereotopic ligands at a prochiral centre (diaslereosmoers)

Conformation of Molecules

In general, there is free rotation about a carbon-carbon and other single *(σ)* covalent bonds. In ethane, for example, the two methyl groups are able to turn freely about the axis between them and there is an infinite number of positions in which the hydrogen atoms on one carbon atom can be arranged relative to those of the second methyl group. Different arrangements of atoms within a molecule whose interconversion does not involve bond fission are known as *conformations*. The two extreme conformations in the ethane molecule – the *eclipsed* and *staggered conformations* – are shown in Figs. 32 *(a)* and *(c)*; Figs. 32 *(b)* and *(d)* show the two conformations of the molecule viewed end-on along the line of the carbon-carbon bond.

The two conformations do not have the same energy. The hydrogen atoms in the two groups tend to repel one another and the staggered conformation, in which the separation of the hydrogen atoms is a maximum, is the more stable form. As the groups are rotated through an angle of 60° to the eclipsed form the potential energy increases to a maximum (see Fig. 33). The energy difference is approximately 12 kJ mol^{-1}. This is too small to allow the two conformations to be isolated as the energy from the thermal collisions of the molecules is sufficient to overcome this barrier and bring about interconversion of the two forms. Rotation about carbon-carbon single bonds is hindered, however, by the presence of large substituents on neighbouring groups. For example, two conformational isomers have been detected in 1,1,2,2-tetrabromoethane at low temperatures.

The barrier to rotation about carbon-carbon double bonds, on the other hand, is extremely high (*c*. 170 kJ mol^{-1} in ethene) and is too large to overcome except under very vigorous conditions. Alkenes substituted on different sides of the double bond do not interconvert, but may be obtained as separate *geometrical isomers*.

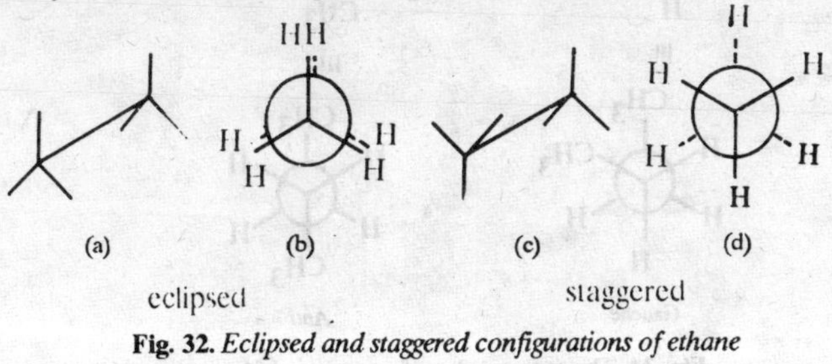

Fig. 32. *Eclipsed and staggered configurations of ethane*

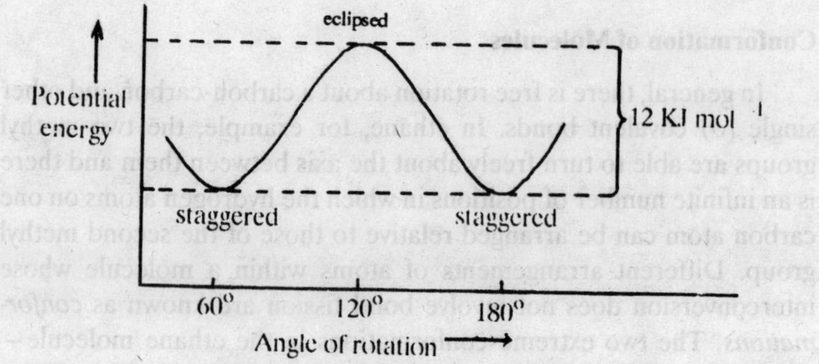

Fig. 33. *Potential energy changes for rotation about the carbon-carbon bond of ethane.*

different sides of the double bond

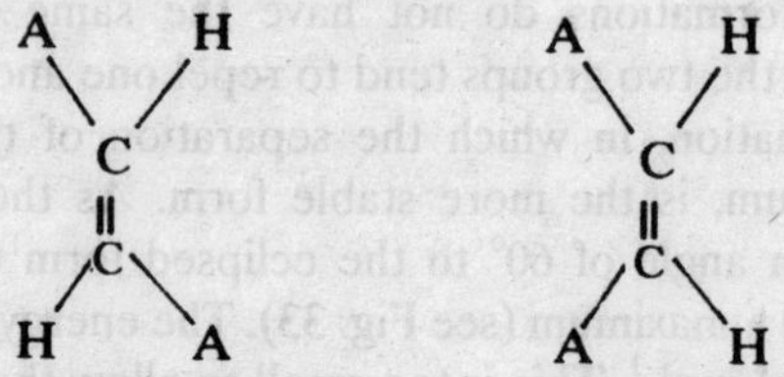

trans-isomer *cis*-isomer

do not interconvert, but may be obtained as separate *geometrical isomers.*

In the case of the butane molecule there are two possible staggered conformations. They are called the *gauche* and *anti* con-

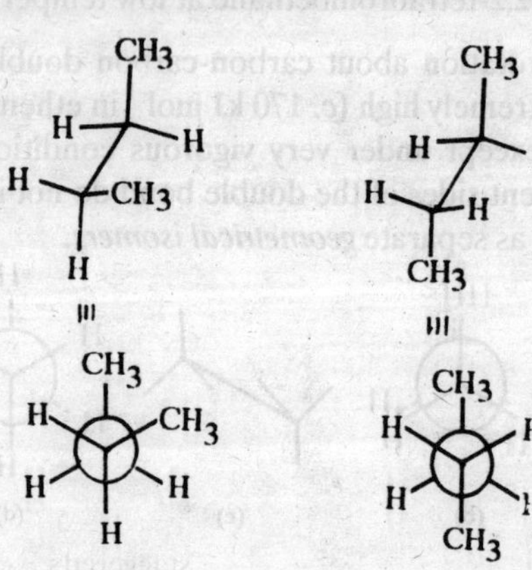

Gauche *Anti*

Fig. 34. The staggered conformations of butane.

formations (Fig. 32). These are conveniently represented by means of *Newman projection formulas* in which the molecule is viewed end-on so that the two carbon atoms are seen one behind the other.

The *gauche* and *anti* conformations are the most stable conformations of the butane molecule, the *anti* conformation being slightly more stable than the *gauche* because the two methyl groups are as far apart as possible. However, although there is a significant energy barrier associated with rotation about the central carbon-carbon bond, it is not large enough to prevent rapid interconversion of the conformers at room temperature. The differeing energy content of the conformations does mean, however, that at any instant 69% of the butane molecules possess the *anti* conformation and 31% the *gauche* conformation.

Energy Differences between Different Conformations

The energy of various conformations of the ethane molecule can be plotted against the torsional angle and it has been shown that this follows a sinusoidal curve (Fig. 35). The system has maximum energy when the hydrogen atoms are eclipsed (torsional angle = 0, 120, 240°) and the minimum energy when the hydrogen atoms are staggered (torsional angle = 60, 180, 300°). The difference between these two energy levels is about 12 kJ mol^{-1} and is too small to permit separation of isomeric molecules having eclipsed and staggered arrangements of the hydrogen atoms. The lower stability of the system with eclipsed atoms is believed to arise from interaction of the electron clouds which constitute the C-H bonds. This is referred to as **torsional energy or torsional strain.**

The position is a little more complex with butane since there are two staggered arrangements each with lower energy than the two

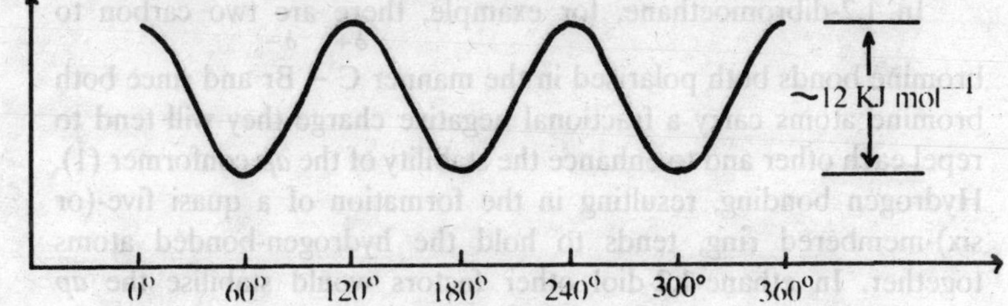

Fig. 35. Ethane

eclipsed arrangements. The difference between the two conformers with staggered groups and the difference between the two conformers with eclipsed groups result from **steric interaction** between non-bonded groups (van der Waal's forces). These are considered to be insignificant between two hydrogen atoms, small between hydrogen and methyl, and larger between two methyl groups. The energy distribution shown in Fig. 36 is consistent with the view that torsional strain is usually greater than steric strain. These energy differences lead to the conclusion that, at 25°, the *ap* (69.0 per cent) and *sc* (30.4 percent) forms predominate.

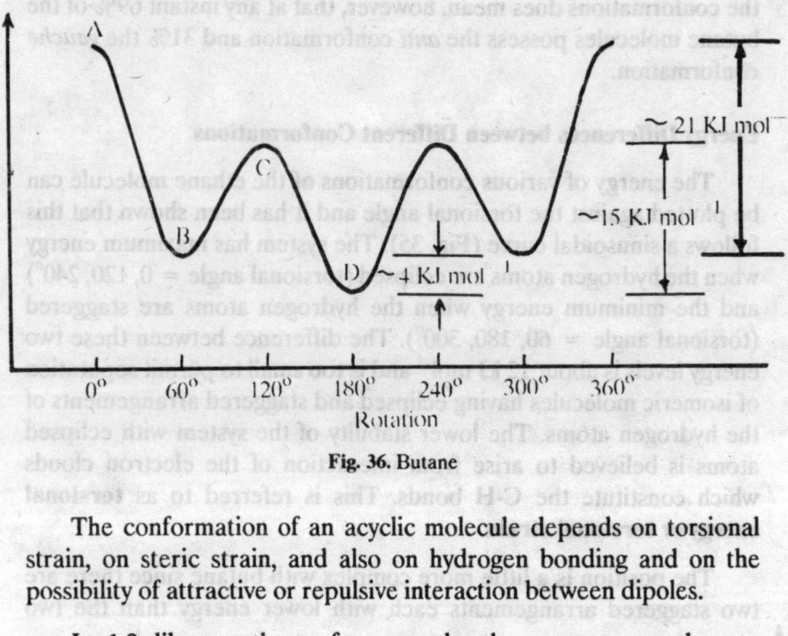

Fig. 36. Butane

The conformation of an acyclic molecule depends on torsional strain, on steric strain, and also on hydrogen bonding and on the possibility of attractive or repulsive interaction between dipoles.

In 1,2-dibromoethane, for example, there are two carbon to bromine bonds both polarised in the manner $\overset{\delta+}{C} - \overset{\delta-}{Br}$ and since both bromine atoms carry a fractional negative charge they will tend to repel each other and to enhance the stability of the *ap* conformer (1). Hydrogen bonding, resulting in the formation of a quasi five-(or six)-membered ring, tends to hold the hydrogen-bonded atoms together. In ethane-1,2-diol other factors would stabilise the *ap* conformer (2) but hydrogen bonding stabilises the *sc* conformation (3).

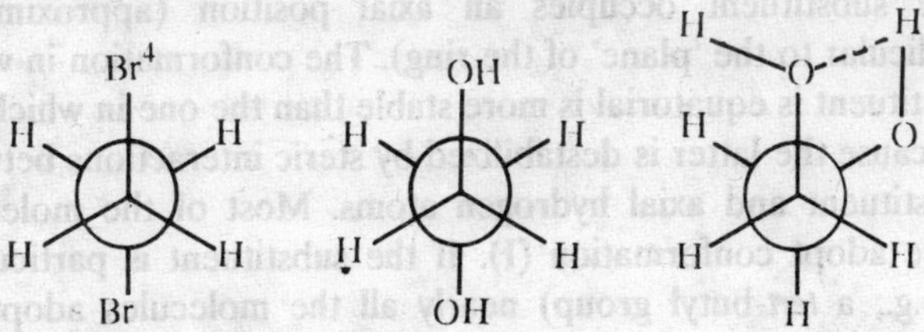

1(*ap*) 2(*ap*) 3(*sc*)

Conformations of Cyclohexane Derivatives

Whereas open chain molecules are very flexible, small cyclic molecules do not posses the same degree of flexibility. Cyclohexane molecules, for example, can adopt either 'chair' or 'boat' conformations which are easily interconverted (Fig. 35). The 'boat' conformation drawn in Fig. 35 is an idealized one. In reality it is slightly twisted and is more correctly referred to as a 'skew-boat'. (It will be useful if you make a model of the cyclohexane molecule to demonstrate that these are strain-free structures and can be interconverted).

The chair conformation is more stable than the boat because the steric interactions between hydrogen atoms are less. In practice, therefore, nearly all cyclohexane molecules adopt chair conformations. The variation in energy content as the conformations are interconverted can be depicted by means of an energy diagram (Fig. 36).

For cyclohexane, itself there are two equivalent chair conformations. However if the cyclohexane ring carries a substituent then the two chair conformations are no longer equivalent (Fig. 37). In one conformation (I) the substituent occupies an *equatorial* position (i.e.,

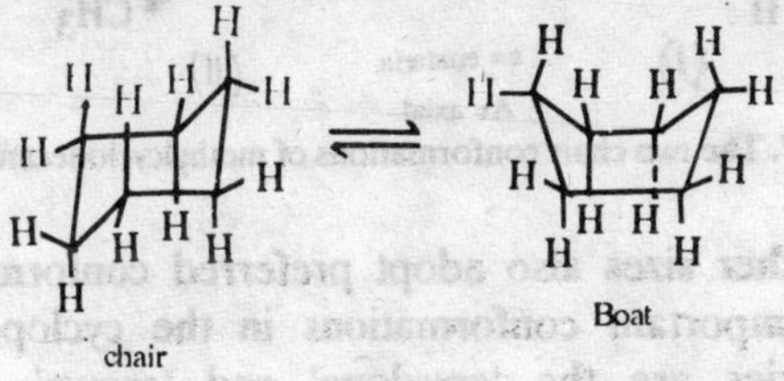

Fig. 35. The two conformations of cyclohexane

approximately parallel to the 'plane' of the ring) whereas in the other (II) the substituent occupies an axial position (approximately perpendicular to the 'plane' of the ring). The conformation in which the substituent is equatorial is more stable than the one in which it is axial because the latter is destabilized by steric interactions between the substituent and axial hydrogen atoms. Most of the molecules therefore adopt conformation (I). If the substituent is particularly large (e.g., a *tert*-butyl group) nearly all the molecules adopt the conformation in which the substituent is equatorial.

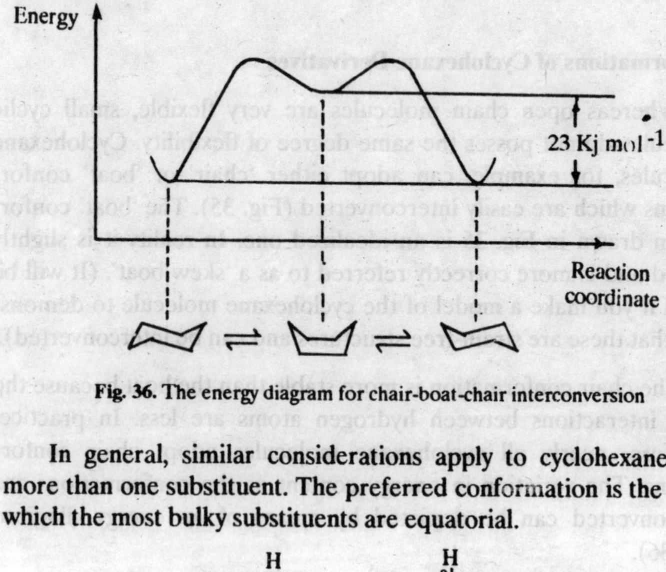

Fig. 36. The energy diagram for chair-boat-chair interconversion

In general, similar considerations apply to cyclohexanes with more than one substituent. The preferred conformation is the one in which the most bulky substituents are equatorial.

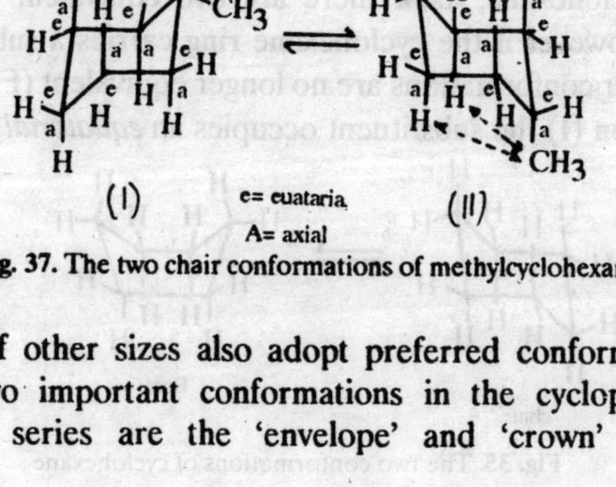

(I) (II)

Fig. 37. The two chair conformations of methylcyclohexane

Rings of other sizes also adopt preferred conformations. For example, two important conformations in the cyclopentane and cyclooctane series are the 'envelope' and 'crown' respectively (Fig. 38).

Fig. 38. The 'envelope' and 'crown' conformations of cyclopentane and cyclooctane

Substituted Cyclohexanes

With substituents larger than hydrogen atoms it becomes significant that the equatorial positions are further apart than the axial positions and in a monosubstitued cyclohexane the equatorially substituted conformer is usually more stable than the axially substituted conformer. In other words, the molecule will spend more of its time in that conformation which places the substituent in an equatorial position. This disparity becomes more marked with increasing size of the substituent and the bulky t-butyl group is used to "lock" a cyclohexane ring in the chair conformation in which the t-butyl is equatorial.

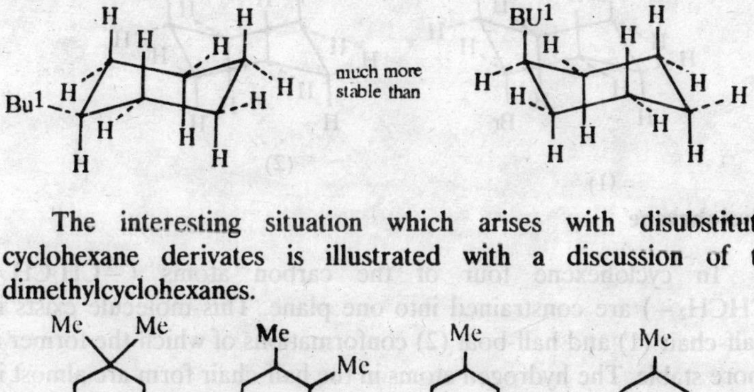

The interesting situation which arises with disubstituted cyclohexane derivates is illustrated with a discussion of the dimethylcyclohexanes.

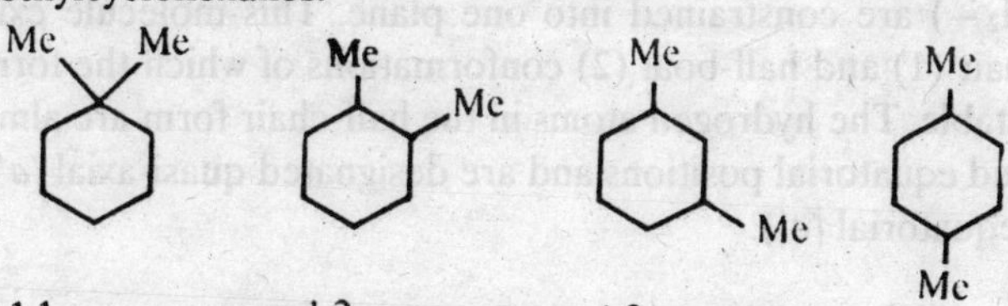

1.1 1,2- 1,3- 1,4-

dimethylcyclohexanes

1,1-Dimethylcyclohexane exists only in one form and does not interest us hęre. The 1,2-, 1,3- and 1,4-dimethyl compounds each exist as *cis*- and *trans*-isomers and it is interesting to enquire (i) what are the possible conformations of each of the six isomers ? (ii) for each isomer which is the more stable conformation ? and (iii) for each dimethylcyclohexane is the *cis*- or the *trans*-isomer the more stable ? There are two conformations for each of the six isomers and sometimes they are equivalent : where they are not the more stable conformer will contribute more to the structure under consideration.

This type of study – known as **conformational analysis** – leads, even in this relatively simple case, to the unexpected conclusion that *cis*-1,3-dimethylcyclohexane is more stable than its *trans*-isomer. Experimental observations support this conclusion.

Other factors sometimes intervene and lead to a different conclusion. Two interesting examples are cited. In *trans*-1,2-dibromocyclohexane the diaxial conformation (1) makes a more significant contribution because in this form, and not in the diequatorial conformation, the carbon-bromine dipoles are opposed to each other. *cis*-Cyclohexane-1,3-diol is stabilised through being strongly hydrogen bonded and this is only possible when the hydroxyl groups are diaxial (1).

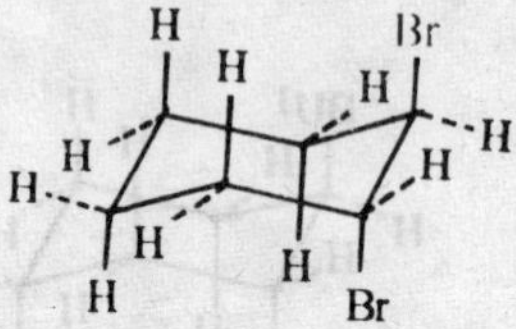

(1)

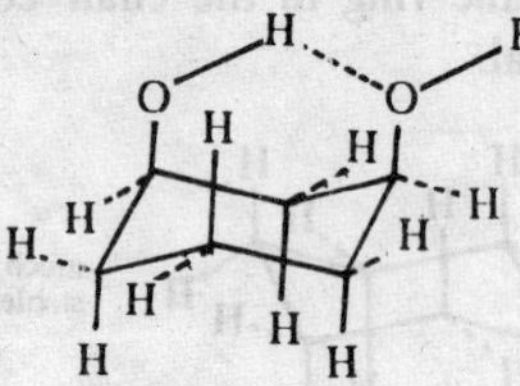

(2)

Cyclohexene

In cyclohexene four of the carbon atoms ($-CH_2CH=CHCH_2-$) are constrained into one plane. This molecule exists in half-chair (1) and half-boat (2) conformations of which the former is more stable. The hydrogen atoms in the half-chair form are almost in axial and equatorial positions and are designated quasi-axial *(a′)* and quasi-equatorial *(e′)*.

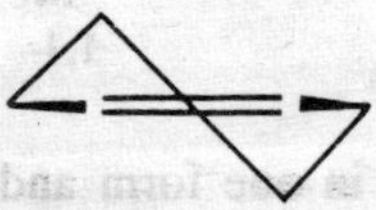

(1)

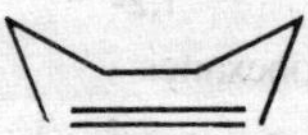

(2)

Decalin and other Polycyclic Systems

Decalin (decahydronaphthalene 1), has a molecular structure corresponding to two fused cyclohexane rings and may be considered as a derivative of 1,2-methylcyclohexane (2). Both these compounds exist in *cis* and *trans* forms and those for decalin are represented by structures 3 *(trans)* and 4 *(cis)*. Six-membered rings are too small to permit di-axial ring fusion and therefore *trans*-decalin exists in only one double-chair conformation which is di-equatorially fused. *cis*-Decalin exists in two enantiomeric conformatios which interconvert easily through a two-boat intermediate.

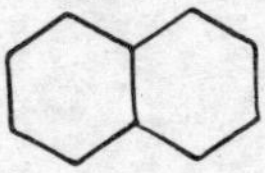

(1)

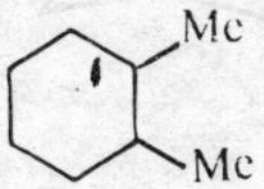

(2)

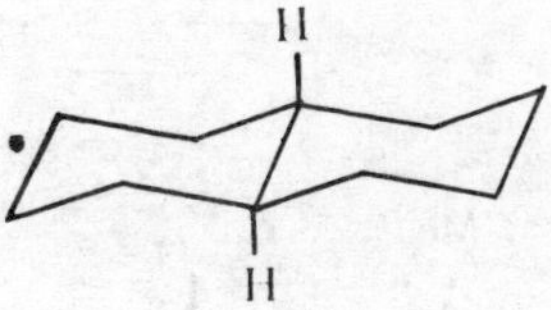

(3) *trans*

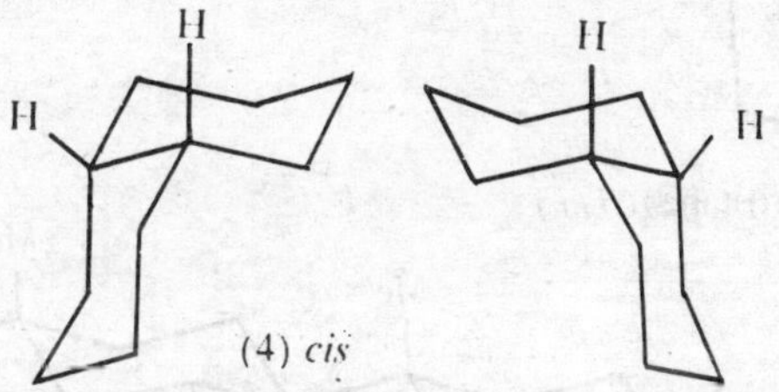

(4) *cis*

It is a short step from decalin to important natural products, such as the steroids, which are based on **perhydro**-1,2-cyclopentanophenanthrene (5). The three six-membered rings exist in chair conformations but are linked together in a rigid structure with axial and equatorial groups clearly defined. The appreciation of this has contributed greatly to our understanding of the chemical reactions of these compounds. There are two series of steroids based on the saturated compounds coprostane and cholestane. These differ only in the mode of linking between rings A and B which is *trans* in cholestane (a derivative of *trans*-decalin) and *cis* in coprostane (a derivative of *cis*-decalin). The remaining rings are *trans* linked.

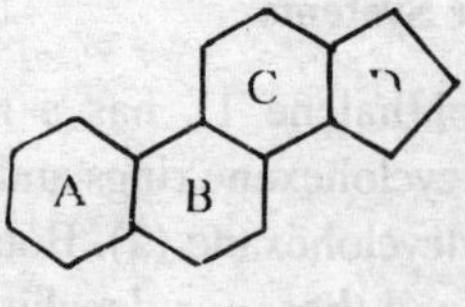

(5)

Cholesterol, the most common of all sterols, is a derivative of cholestane with the structure shown above.

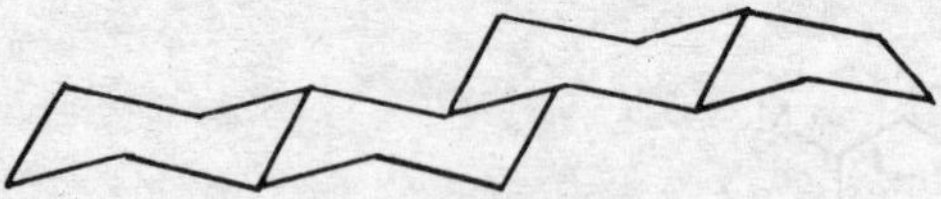

cholestane (*t.t.t.t*)

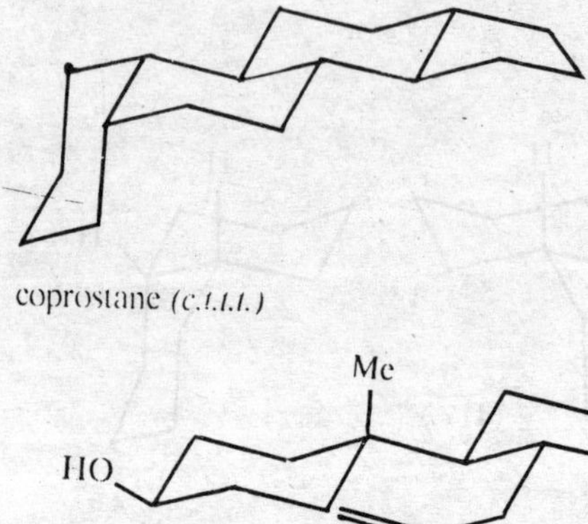

coprostane (*c.t.t.t.*)

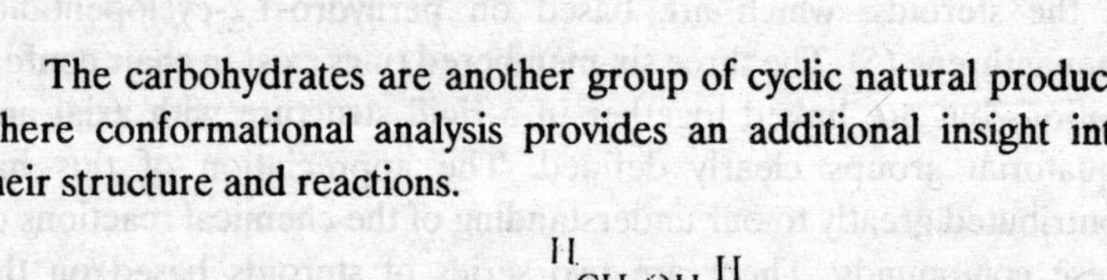

Cholesterol

The carbohydrates are another group of cyclic natural products where conformational analysis provides an additional insight into their structure and reactions.

a-D-glucopyranose

Influence of Conformation on Chemical Properties

Groups in axial and equatorial positions differ in their thermodynamic stability and in their steric accessibility and these factors influence the chemical properties of molecules in differing conformations.

Equilibrium reactions will furnish the more stable isomer. For example, when ethyl 4-*t*-butylcyclohexanecarboxylate is heated with sodium ethoxide both the *cis*- and *trans*-isomers produce the same equilibrium mixture containing about 84 per cent of the more stable *trans*-isomer.

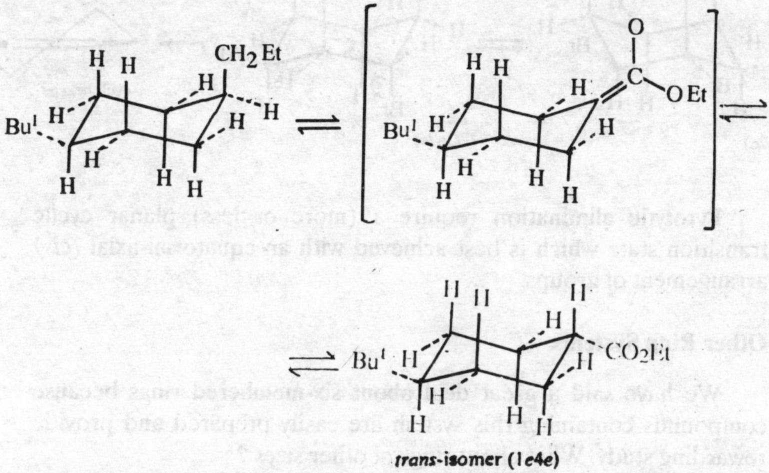

trans-isomer (1e4e)

In kinetically controlled reactions consideration has to be given to the steric requirement of the transition state and to whether the reaction is controlled by its formation or its further reaction. In reactions in which the formation of the transition state is rate-controlling equatorial substituents are more reactive than similar axial substituents because of the lower steric restriction to the formation of the transition state in the equatorial position. The esterification of cyclohexanols and cyclohexanecarboxylic acids occurs more rapidly when the hydroxyl or carboxyl group is equatorial. The position is often reversed (as in the oxidation of cyclohexanols by chromic acid) when reaction is controlled by the decomposition of a transition state or intermediate because of the greater possibility of the relief of steric congestion, leading to steric acceleration in the axial position. In bimolecular substitution reactions axial substituents are usually displaced more readily than equatorial substituents because the rearward approach of the attacking reagent is more hindered for equatorial than for axial substituents.

Elimination reactions are most commonly *trans* and require an antiperiplanar relationship between the two groups to be eliminated. This is only possible in the diaxial conformation and therefore *trans*-isomers, which can assume a diaxial conformation, react more readily than *cis*-isomers, which cannot.

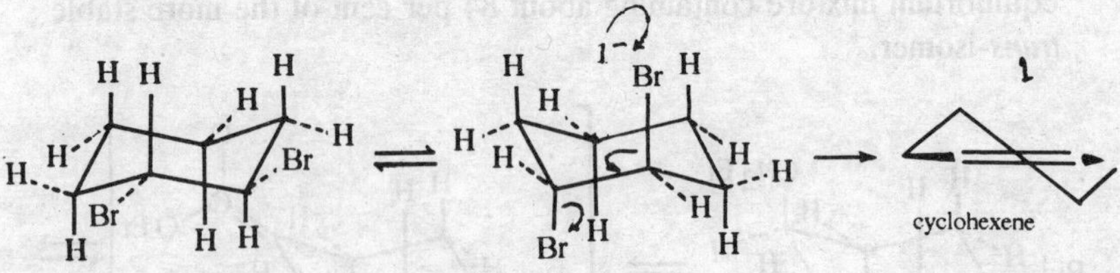

(1e2e) (1a2a)

Pyrolytic elimination require a (more or less) planar cyclic transition state which is best achieved with an equatorial-axial *(cis)* arrangement of groups.

Other Ring Systems

We have said a great deal about six-membered rings because compounds containing this system are easily prepared and provide rewarding study. What about rings of other sizes ?

The stability of a ring system depends on several factors and the most stable form of the ring will be that in which these are balanced to give the minimum energy. The factors to be considered are :

(i) Angle strain resulting from the deviation from the normal tetrahedral angle (109°28′).

(ii) Torsional strain resulting from non-bonded interactions between groups attached to adjacent atoms.

(iii) Steric strains resulting from non-bonded interactions between groups on non-adjacent atoms as, for example, between 1,3-diaxial substituents in cyclohexane derivatives. This becomes more significant in medium sized rings where transfer of groups across a ring is frequently observed (transannular reactions).

The ease of formation of a ring system will also depend on an additional factor. Most synthetic procedures involve an intramolecular reaction between two reactive centres in a single molecule and the ease of this reaction depends, in part, on how much of their time the two reactive centres spend in close propinquity. This will decrease with increasing separation of the two reaction centres since the number of possible conformations which an acyclic system may assume will increase with its length. It is for this reason that three-membered rings, though subject to considerable strain, are nevertheless formed more easily than four-membered rings.

Four- and five-membered rings are not planar since puckering results in some reduction of torsional strain. Rings larger than cyclohexane are, like this molecule, flexed to reduce both angle strain and torsional strain though this may result in significant transannular interactions. Possible conformations for cyclo-octane and cyclododecane are shown.

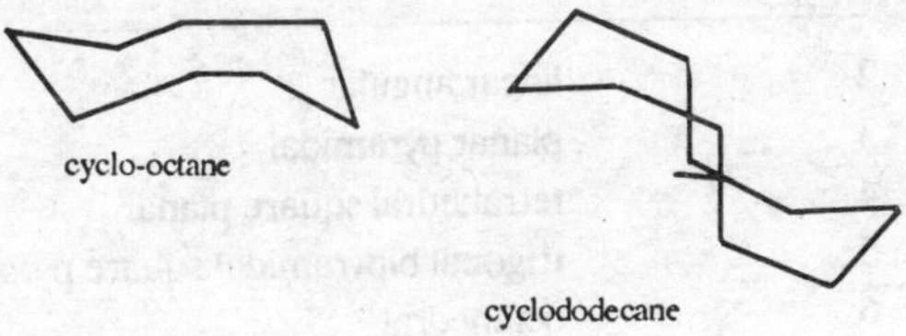

Stereochemistry of compounds of elements other than carbon

So far in this book attention has been focussed almost entirely on organic compounds. This situation arises from the facts that stereoisomerism is widely observed among organic compounds, that its nature has been recognised for over a century, and that methods of producing chiral molecules are most highly developed among carbon compounds. Nevertheless, stereochemistry is not confined to carbon compounds and the discussion will now be extended to other elements.

We begin with some general comments on molecular geometry, *i.e.*, the spatial disposition of ligands around a central atom. The atom

A may be attached to several ligands to give compounds AL_n and such compounds may also contain lone pairs E associated with A which, along with the bonding electron pairs, influence molecular geometry. These compounds should be considered as AL_nE_m. Molecular geometry depends on the number of valence shell electron pairs (bonding *and* lone) possessed by A.

The molecular shapes which a compound AL_n may assume are set out in Table 2 for values of *n* between 2 and 8. The previous account of the stereochemistry of organic compounds is really an extended discussion of one section of Table 2, *viz.*, compounds in which A has a coordination number of four and in which the ligands assume a tetrahedral distribution around A. We now examine some other systems and consider the possibilities of stereoisomerism and the stability of stereoisomers. It will be recalled that a compound is potentially enantiomeric if its molecular structure cannot be superimposed on its mirror image and that stable enantiomers will exist only if there is an appropriate energy barrier between the two forms.

Table 2

Number of ligands	*Shape of molecule*
2	linear angular
3	planar pyramidal
4	tetrahedral square planar
5	trigonal bipyramidal square pyramidal
6	octahedral
7	pentagonal pyramidal
8	cubic square antiprismatic dodecahedral, etc.

Carbon is not the only atom with a coordination number of four and a tetrahedral arrangement of ligands. Other elements share these properties and the possibilities of chirality, discussed at length in Chapter 3, apply equally to appropriate derivatives of other elements. It is not surprising, therefore, that optically active compounds of silicon are known. Derivatives of Group 5 elements which can be written as AL_3E or AL_4 are, however, of greater interest. In both of these classes of compounds the central atom is linked to four groups – one of which may be a lone pair – and if these are tetrahedrally arranged then there should be some similarity between these and analogous carbon compounds in respect to stereochemical possibilities. It is thus not surprising that ammonium and phosphonium salts exists in non-superimposable (chiral) forms and the oxide 1 and salts of the ions 2 and 3 in which nitrogen or phosphorus is attached to four different ligands have been separated into enantiomeric forms.

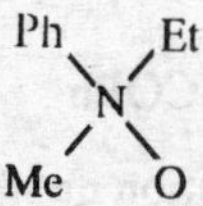

(1)

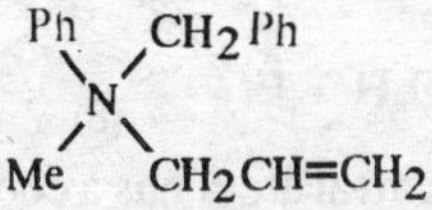

(2)

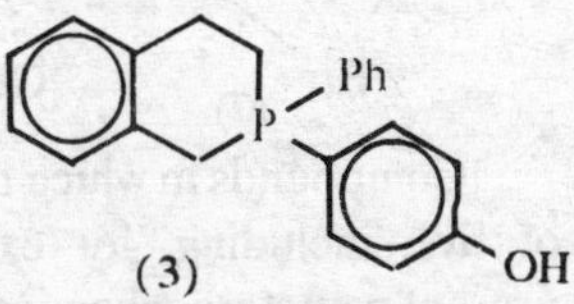

(3)

Trivalent derivatives of nitrogen, phosphorus, arsenic, and sulphur exist in pyramidal form though this can be considered as tetrahedral if the lone pair is counted as a fourth ligand. Among nitrogen derivatives inversion of pyramidal forms occurs very easily via a planar transition state so that the separation of enantiomers (4) and (5) is not possible except in special cases.

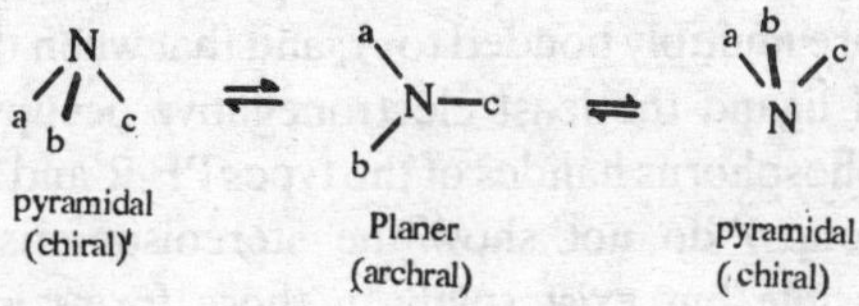

(4) (5)

The ammonia molecule is known to be pyramidal with a height of 0.30 A but inversion occurs at a rate of 2.4×10^{10} cycles per second, the energy of activation for this change being only 25 J mol^{-1}. This reversible inversion can be prevented by holding the nitrogen pyramid in a rigid system and compound (6) in which both nitrogen atoms are part of two ring systems has been isolated in an optically active form.

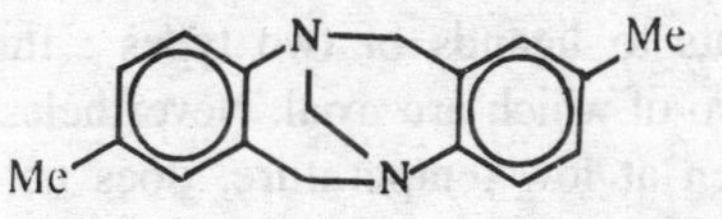

(6)

Phosphorus, arsenic, and sulphur derivatives have a higher energy of activation and the rate is slow enough to permit separation of diastereoisomeric camphorsulphonates of the sulphonium ions (7) and (8).

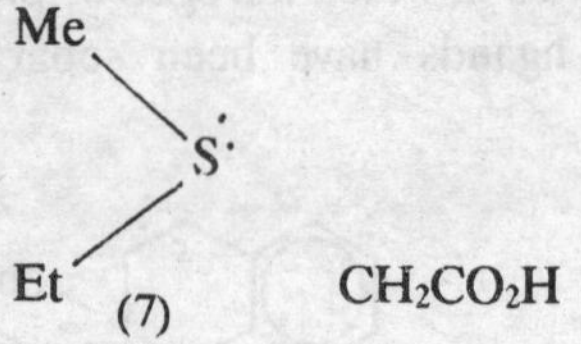

(7)

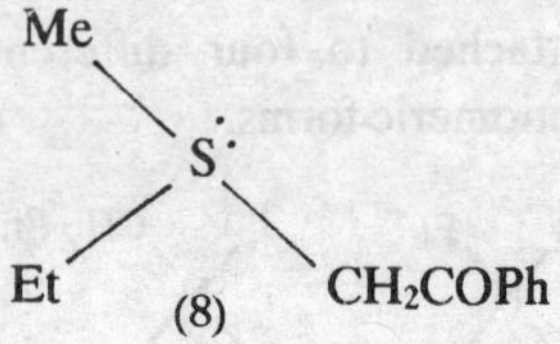

(8)

Compounds in which the central atom has a coordination number of five, including, for example, many derivatives of phorphorus, present new stereochemical features. Such compounds exist mainly as trigonal bipyramids and less commonly as square pyramids. Spectroscopic studies show that compounds like PF_5 or $Fe(CO)_5$ exist as trigonal bipyramids : three of the ligands (L_e, equatorial) are in the same plane as the central atom and attached to it by bonds at angles of 120° and the remaining two ligands (L_a, axial, apical) are positioned above and below the equatorial plane. It is now accepted that molecules of the type AL_4E, AL_3E_2, and AL_2E_3 will have shapes based on a trigonal bipyramid, that lone pairs occupy equatorial positions, as do atoms which are multiply bonded to A, and that when there is more than one type of ligand the least electronegative occupy equatorial positions. Thus phosphorus halides of the types PF_4R and PF_3R_2 (R = alkyl or aryl groups) do not show the stereoisomerism which is theoretically possible but exist solely in those forms in which the organic groups occupy equatorial positions (45 and 46).

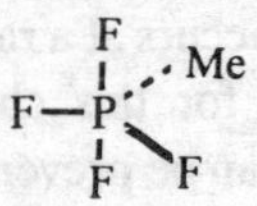

(9)

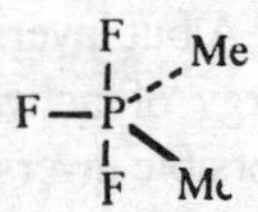

(10)

All this points to ligands of two types : three of which are equatorial and two of which are axial. Nevertheless the careful nmr study of PF_5, even at low temperature, does not give two signals corresponding to axial and equatorial fluorine atoms. The compound Me_2PF_3, however, does show two ^{19}F signals at low temperature. The

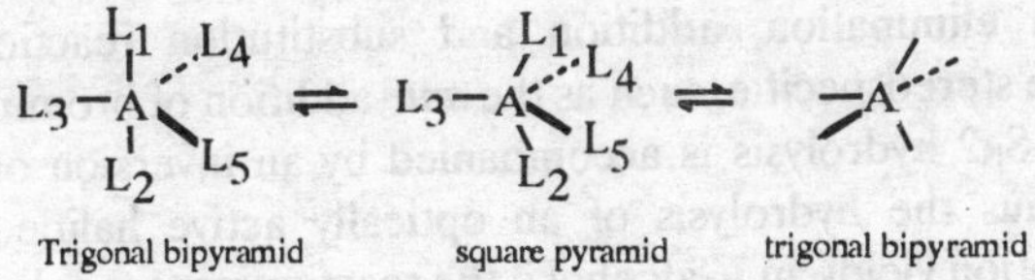

unexpected result for PF_5 is usually discussed in terms of stereochemically non-rigid structures; a term applied to molecules that undergo rapid intramolecular rearrangement. This facile internal rearrangement of the trigonal bipyramid has been explained by a number of mechanisms and one that finds common acceptance is pseudorotation first proposed by Berry. If the L_1AL_2 angle, normally 180°, is reduced slightly and the L_4AL_5 angle, normally 120°, is increased slightly then a square pyramid results and this can either revert to the original trigonal bipyramid or change to a second trigonal bipyramid. The consequence of this inversion is that two equatorial ligands have become axial and two axial ligands have become equatorial. The equatorial bond not involved in this change is described as the "pivot bond" but since any two of the three equatorial ligands are involved in pseudorotation, all five ligands become equivalent. The problem is more complex when pairs of ligands are part of the same cyclic system.

This non-rigidity means that even in apparently chiral structures, enantiomeric compounds are unlikely to be isolated.

Compounds of atoms with a coordination number of six show octahedral geometry. The octahedron, like the tetrahedron, is a stable (rigid) configuration and many compounds are chiral and are known in their enantiomeric forms. Examples are to be found among the enantiomeric metal complexes such as 11 and 12.

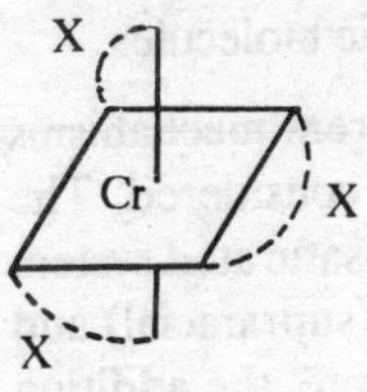

(11)

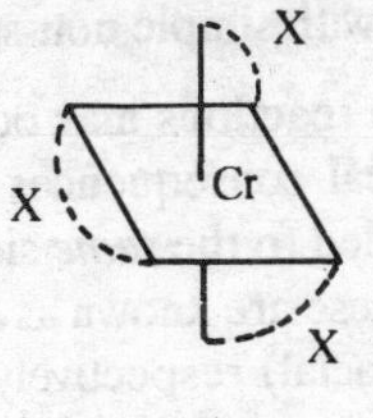

(12)

X = $NH_2CH_2CH_2NH_2$ or —OOCCOO—

Stereospecific Reactions

Certain elimination, addition and substitution reactions are known to be stereospecific, such as the *anti*-addition of bromine to an alkene. An S_N2 hydrolysis is accompanied by an inversion of configuration; thus the hydrolysis of an optically active halide with a D-configuration yields an L-alcohol if the reaction proceeds by an S_N2 mechanism :

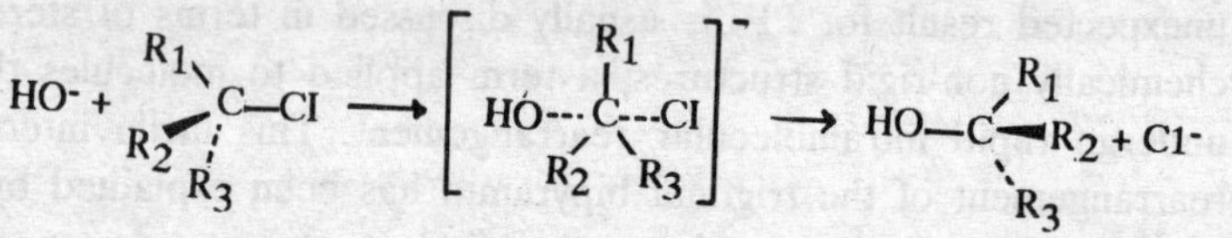

An S_N1 mechanism results in racemization. Any reaction in which the configuration of a compound is inverted in this way is known as a *Walden inversion*.

Stereochemistry of Addition Reactions

Many important reactions of alkenes and alkynes are addition processes which can be written in a general form such as :

$$\underset{(1)}{RC \equiv CR'} \xrightarrow{X_2} \underset{(2)}{RCX = CXR'}$$

$$\underset{(3)}{RCH = CHR} \xrightarrow{X_2} \underset{(4)}{RCHXCHXR'}$$

The alkyne (3) exists in only one form but the alkenes (2 and 3) exist in *Z* or *E (cis* and *trans)* forms and the addition product (4), with two non-identical chiral centres, exists in four enantiomeric forms which can be paired into two racemates *(threo* and *erythro)*.

Addition reactions which are not stereospecific are of little synthetic value except with simple non-stereoisomeric molecules.

Although addition reactions may occur by different mechanisms only two stereochemical consequences have to be considered. The two groups may be added to the *same* side of the unsaturated system or to *opposite* sides. These are known as *cis* addition (suprafacial) and *trans* addition (antarafacial) respectively. The substrate, the addition process, and the products are then related as shown in Scheme 1.

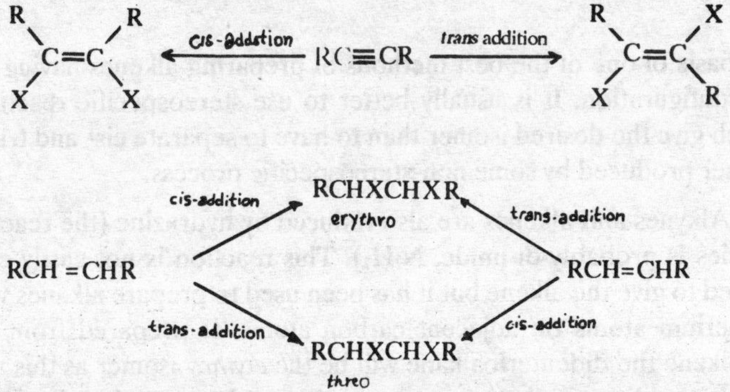

Scheme 1. Stereochemical relationships between alkynes and alkenes and their addition products.

The statements formulated in these sequences will be established shortly but first we examine their significance. An *erythro* adduct results either from a *cis*-alkene by *cis*-addition or from a *trans*-alkene by *trans*-addition. Conversely the *threo* adduct results from the *cis*-alkene by *trans*-addition or from the *trans*-alkene by *cis*-addition. These observations are significant when there are two ways of adding X_2 to an alkene, one involving *cis*-addition and the other involving *trans*-addition. If there is only one procedure available then a particular adduct can only be obtained from the appropriate alkene. For example, if only a *cis*-addition procedure is known then the *erythro* adduct must be obtained from the *cis*-alkene and the *threo* adduct must be obtained from the *trans*-alkene. Obviously this type of information must be fully appreciated before embarking on a synthetic sequence.

In general, *cis*-addition may occur equally well from below the molecule. Show by a similar sequence that this gives the other *erythro* enantiomer so that the addition product is the racemic *erythro* form.

Steroechemistry of Hydrogenation

Catalytic hydrogenation is usually a *cis*-addition and hydrogenation of an alkyne proceeds via the *cis*-alkene to the alkane which

$$RC \equiv CR' \xrightarrow{H_2Pd} RCH = CHR \xrightarrow{H_2Pd} RCH_2CH_2R'$$

is not stereoisomeric. The reaction can be halted at the *cis*-alkene stage if a partially poisoned palladium catalyst is employed and this is the basis of one of the best methods of preparing alkenes having the *cis*-configuration. It is usually better to use stereospecific reactions which give the desired isomer than to have to separate cis- and trans-isomer produced by some non-stereospecific process.

Alkynes and alkenes are also reduced by hydrazine (the reactive species is probably di-imide, N_2H_2). This reaction is not easily controlled to give the alkene but it has been used to prepare alkanes with deuterium atoms on adjacent carbon atoms. If prepared from the *cis*-alkene the dideuterioalkane will be the *erythro* isomer as this also is a *cis*-addition, reaction occurring through a quasi-cyclic intermediate. Conversely the *threo* adduct can be prepared from the *trans*-alkene.

$$RC \equiv CR' \xrightarrow{N_2H_4} RCH = CHR' \xrightarrow{N_2H_4} RCH_2CH_2R$$

$$RCH = CHR' \xrightarrow{N_2D_4} RCHDCHDR'$$

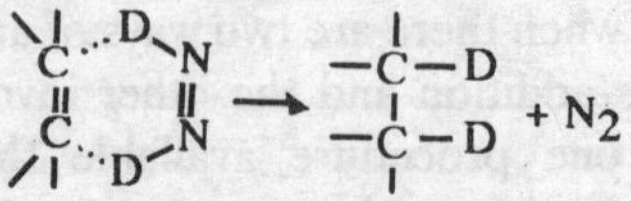

Stereochemistry of Bromination and Hydroxylation

Other important addition reactions include halogenation (especially bromination), which is usually a *trans*-addition, and hydroxylation, which may be *cis* or *trans* depending on the choice of reagent. In summary, halogenation and hydroxylation with peracids are *trans*-addition processes and hydroxylation with potassium permanganate or with osmium tetroxide are *cis*-addition processes.

$$RCHBrCHBrR \xleftarrow{\text{bromination}} RCH = CHR \xrightarrow{\text{hydroxylation}} RCH(OH)CH(OH)R$$

All of these reactions start with an initial *cis*-addition leading to a three- or five-membered ring followed by ring opening which may occur with or without inversion. *cis*-Addition followed by inversion is equivalent overall to *trans*-addition.

Elimination Reactions

Elimination reactions such as dehydration, dehydrohalogenation, dehalogenation, and the elimination reactions shown by quaternary

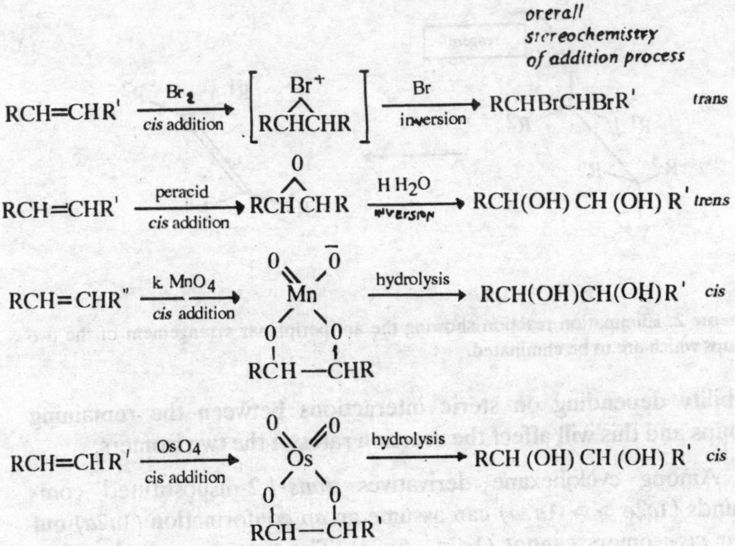

ammonium compounds and certain esters are important for the preparation of alkenes are alkynes. Since both substrate and product may be capable of existing in stereoisomeric forms the stereochemical nature of these reactions becomes significant. In many respects the stereochemistry of elimination processes is the reverse of the stereochemistry of addition processes.

Both radical and polar elimination reactions are known and the latter occur by unimolecular (E1) or bimolecular (E2) mechanisms. The steric nature of the radical and E1 polar processes is difficult to control and to predict but E2 processes are more stereospecific and both *cis* and *trans* elimination processes are known. The latter are more common and will be discussed first.

Trans elimination of two atoms or groups from adjacent (vicinal) carbon atoms normally requires an antiperiplanar arrangement of the two groups which are to be removed. This has consequences for the rate of reaction (and in extreme cases on whether it proceeds or not) and for the stereochemistry of the product.

With acyclic compounds both *threo* and *erythro* isomers can usually assume an *ap* arrangement. These however may vary in

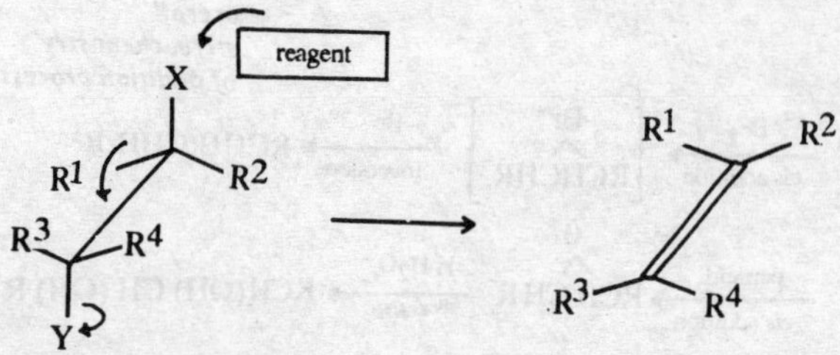

Scheme 2. Elimination reaction showing the antiperiplanar arrangement of the two groups which are to be eliminated.

stability depending on steric interactions between the remaining groups and this will affect the reaction rates of the two isomers.

Among cyclohexane derivatives *trans*-1,2-disubstituted compounds (1*e*2*e* > > 1*a*2*a*) can assume an *ap* conformation *(*1*a*2*a)* but their *cis*-isomers cannot *(*1*e*2*a* - 1*a*2*e*). The *trans*-isomers therefore undergo elimination reactions more readily and in many cases the *cis*-isomers do not react at all or undergo an alternative reaction.

In the conversion of alkenes to alkynes only one isomer can assume the necessary *trans* arrangement of groups.

Dehydrobromination

Dehydrobromination is a *trans*-elimination effected by base and used, for example, in the conversion of dibromides (themselves prepared from alkenes) to alkynes. The bromination- dehydrobromination reaction may be written :

$$RCH = CHR' \xrightarrow[-HBr]{Br2} RCHBrCHBrR \xrightarrow{base} [RCH = CBrR] \xrightarrow[-HBs]{base} RC \equiv CR'$$

There is only one alkyne but two stereoisomeric vinyl bromides and what is known about the addition process (bromination) and the elimination process (dehydrobromination), both of which are *trans*, enables us to decide whether it matters which alkene is used or whether one alkene is superior or perhaps even a requirement for successful synthesis of the alkyne. These possibilities are examined in Schemes 3 and 4, which should be carefully examined. The conclusions are in line with the observation that *cis*-alkenes readily yield alkynes via *threo* dibromides but that *trans*-alkenes do not. The *erythro* dibromides formed from the latter usually give a mixture of allenes and vinyl halides :

$$RCH_2CH = CHCH_2R' \rightarrow RCH_2CHBrCHBrCH_2R \rightarrow \begin{cases} RCH_2CH = CBrCH_2R' \\ RCH_2CBr = CHCH_2R' \\ RCH = C = CHCH_2R' \\ RCH_2CH = C = CHR' \end{cases}$$

trans *erythro*

Dehalogenation

Dehalogenation – most often of dibromides – by reaction with zinc or with sodium iodide is a very useful process. The reaction with iodide ion is more highly stereospecific than that with the metal. Among acyclic compounds *threo* dibromides furnish *cis*-alkenes and *erythro* dibromides give *trans*-alkenes with the *erythro* compound slightly more active since it can more readily assume the antiperiplanar conformation. Steric interaction is somewhat greater in the *ap*-configuration of the *threo* isomer. This point is illustrated in Scheme 5 with one *threo* and one *erythro* enantiomer. The reader should satisfy himself that these observations are equally true for the other enantiomers.

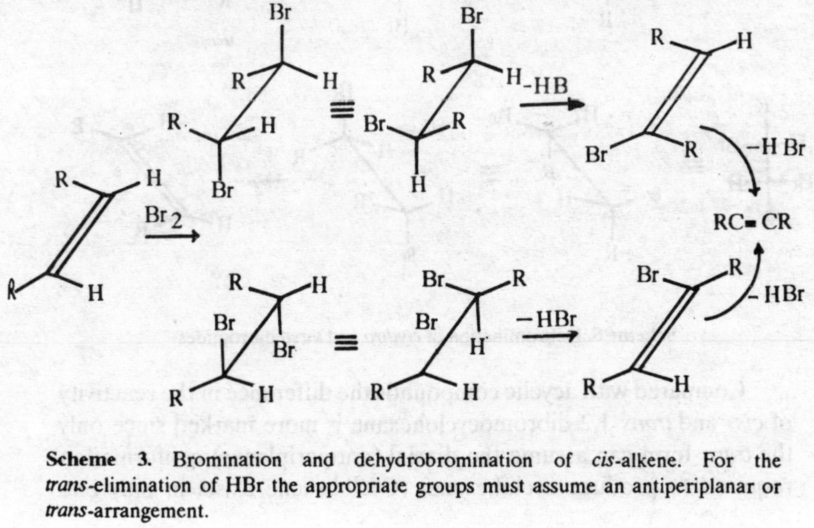

Scheme 3. Bromination and dehydrobromination of *cis*-alkene. For the *trans*-elimination of HBr the appropriate groups must assume an antiperiplanar or *trans*-arrangement.

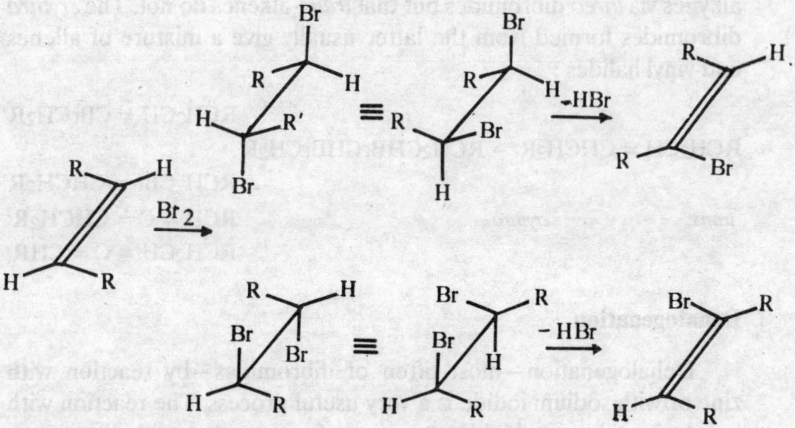

Scheme 4. Bromination and dehydrobromination of a *trans*-alkene. Elimination of one mole of HBr occurs readily from the *erythro* dibromide in an antiperiplanar conformation but the product is a mixture of two vinyl bromides in which bromine and olefinic hydrogen cannot assume a *trans*-arrangement.

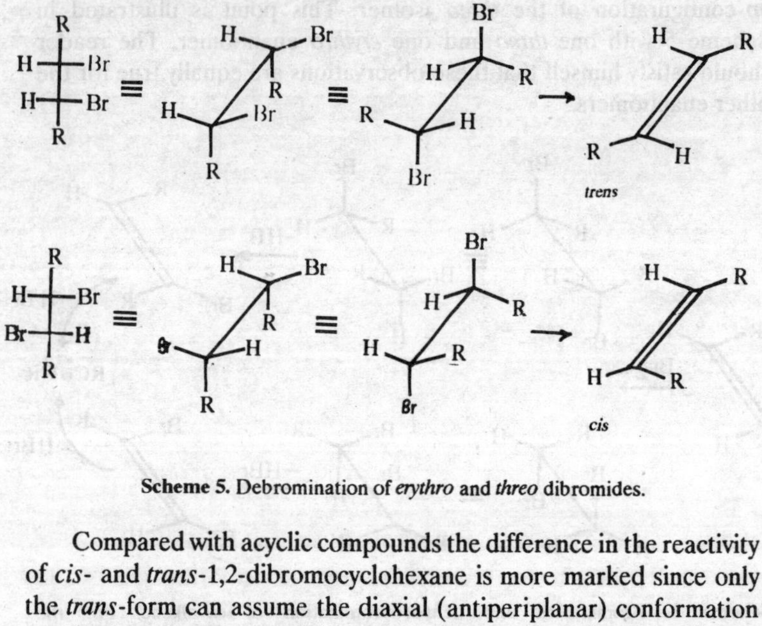

Scheme 5. Debromination of *erythro* and *threo* dibromides.

Compared with acyclic compounds the difference in the reactivity of *cis*- and *trans*-1,2-dibromocyclohexane is more marked since only the *trans*-form can assume the diaxial (antiperiplanar) conformation required to produce cyclohexene. (Cyclohexene exists in only one isomeric form. The smallest ring system in which a *trans* double bond can be accommodated is C_8). The *cis*-bromide forms cyclohexene only slowly, probably after substitution (with inversion) of one of the bromine atoms by iodine. These changes are outlined in Scheme 6.

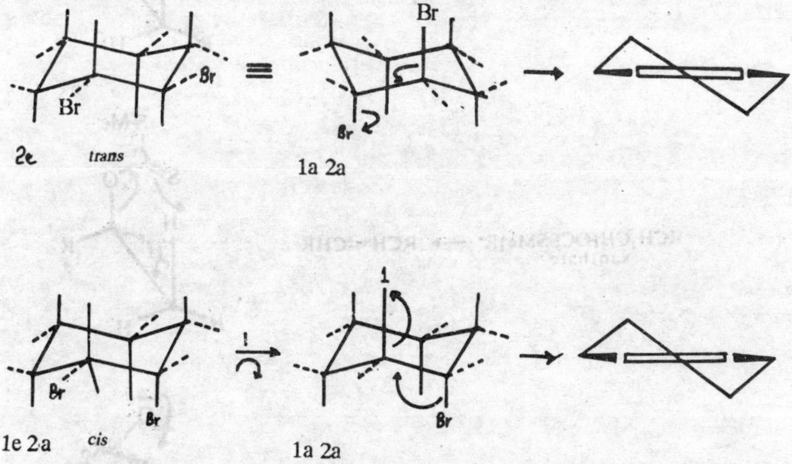

Scheme 6. Debromination of *trans*- and *cis*-1,2- dibromocyclohexane (hydrogen atoms omitted for clarity).

Cis elimination reactions

Important *cis*-elimination reactions occur through intermediates with some cyclic character. These reactions include the pyrolytic cleavage (300-500°) of esters (acetates and xanthates) and of amine-N-oxides. The reaction occurs through a cyclic intermediate (shown to the right of each equation) in which the groups to be eliminated assume the synperiplanar conformation.

An interesting example of *cis*-elimination is found in the pyrolysis of the deuterated compound PhCHDCH(OAc)Ph to stilbene. The acetoxy deuterio compound exists as *threo* and *erythro* isomers only one of which gives a product contianing deuterium. This observation can be rationalised on the basis that in the preferred conformation for each isomer the bulky phenyl groups will be as far away from one another as possible and the *cis*-elimination then involves AcOH or AcOD as appropriate.

$RCH_2CH(OCOCH_3)R' \longrightarrow RCH{=}CHR'$
acetate

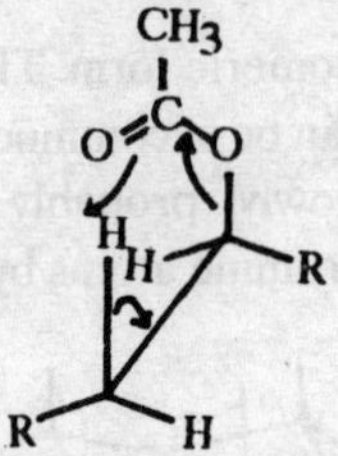

$RCH_2CH(OCSSMe)R' \longrightarrow RCH{=}CHR$
xanthate

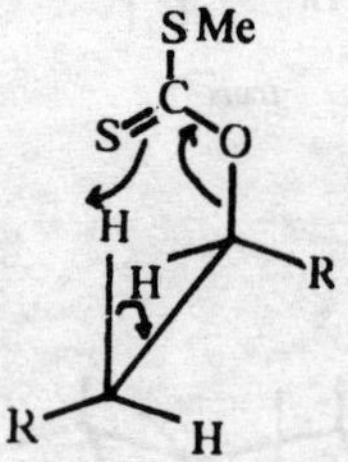

$R'CH_2CH(\overset{+}{N}\overset{-}{O}R_2)R' \longrightarrow RCH{=}CHR'$
amine oxide

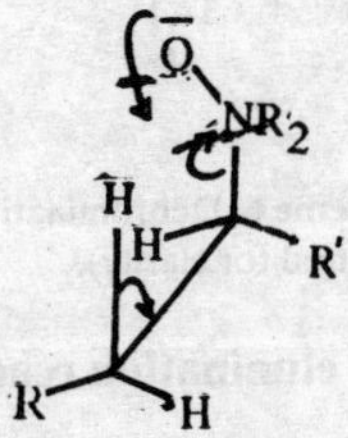

The Corey-Winter synthesis of alkenes from diols via their thionocarbonates is essentially a *cis*-elimination with the *threo* diol furnishing the *trans*-alkene and the *erythro* diol yielding the *cis*-alkene.

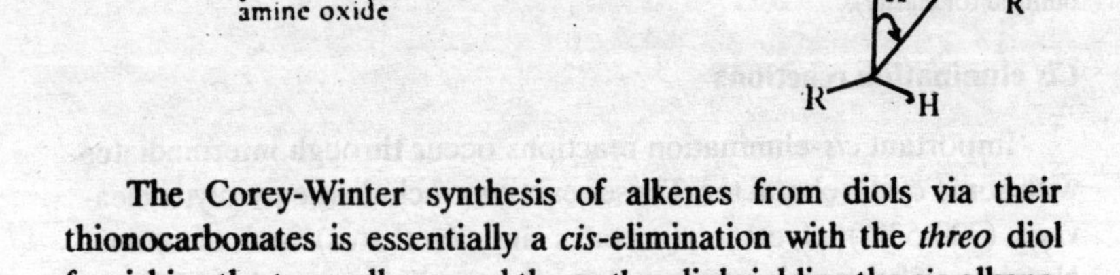

Substitution reactions

It is a useful generalisation, though slightly over-simplified, to say that substitution reactions are accompanied either by inversion or by racemisation. A third possibility—retention of configuration—is

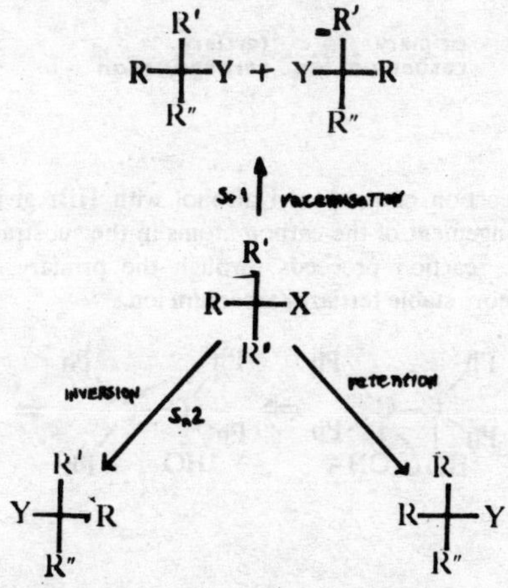

Scheme 7. The possible steric courses of substitution reactions.

relatively uncommon but important when it does occur (Scheme 7). It has been amply demonstrated that bimolecular substitution reactions (S_N2) occur with inversion of configuration and that unimolecular substitution reactions (S_N1) are normally accompanied by racemisation. Retention of configuration is observed in a few S_N1 reactions of compounds containing configuration-holding groups, in S_N1 reactions, and in S_N2 reactions which, on closer examination, are shown to occur through two steps both of which involve inversion.

Molecular rearrangements

In most reactions of organic compounds one functional group is converted to another without any change in the carbon skeleton of the substrate. Those reactions in which this is not true are described as rearrangements. Two familiar examples are formulated in Schemes 8 and 9. There are several interesting stereochemical problems which may arise in rearrangement reactions and three of them are discussed here :

(i) In rearrangements of the type shown in the following equation in which the migrating group R is attached to X through a chiral

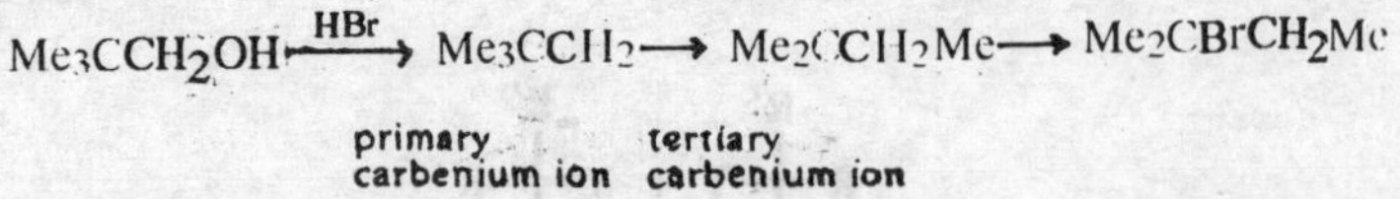

Scheme 8. The reaction of neopentyl alcohol with HBr giving tert amylbromide involves the rearrangement of the carbon atoms in the substrate. This is believed to occur because the reaction proceeds through the primary carbenium ion which rearranges to the more stable tertiary carbenium ion.

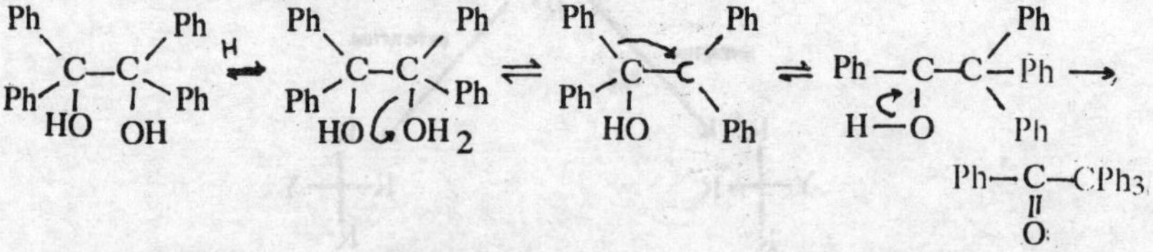

Scheme 9. The pinacol rearrangement : 1,2-diols (especially ditertiary) diols furnish rearranged ketones when treated with acid by the mechanism shown above. This is represented in discrete steps for the sake of clarity but may follow a concerted mechanism.

centre migration may occur with retention of configuration or with racemisation. When reaction occurs with retention of configuration this is taken as evidence of an intramolecular reaction in which the R group is never separated from the molecule; bond-making and bond-breaking occur simultaneously. Racemisation, on the other hand, is associated with intermolcular reactions in which the group R separates from its molecule – most often as a carbenium ion which assumes a planar configuration – and loses its steroechemical identity.

Migration with retention of configuration has been observed in many rearrangement reactions including those associated with the names of Hofmann, Curtius, Lossen, Wolff, Beckmann, and Baeyer and Villiger. The conversion of (+)-3-phenylbutan-2-one to (–)-1-phenylethylacetate by the last-named oxidation is typical (Scheme 10).

(ii) The migration of R from X to Y (equation 1) is usually preceded or followed by the elimination or displacement of some

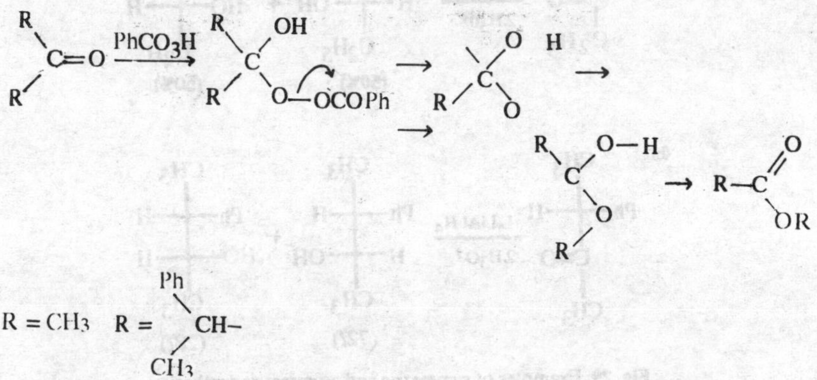

Scheme 10. Baeyer-Villiger oxidation. Migration of R′ from carbon to electron-deficient oxygen occurs without any change of configuration of its chiral centre.

group attached to Y. If Y has four different ligands attached to it then migration will be accompanied by inversion at this centre if this step is bimolecular but by racemisation if the group attached to Y separates first and leaves a carbenium ion which is sufficiently stable to assume a planar configuration.

Asymmetric Synthesis

Any reaction in which a chiral molecule is prepared entirely from achiral or racemic starting materials produces equal amounts of both enantiomers. For example, in the reduction of butanone with $LiAlH_4$ (Fig. 29 (a)) the nucleophilic (AlH_4 can attack either side of the molecule with equal ease, giving equal amounts of both enantiomers. However, if one of the starting materials in a reaction is already chiral then the two product molecules (diastereoisomers) are produced in unequal amounts (Fig. 29(b)) because the two transition states leading to the products are no longer identical. Such reactions are called *asymmetric synthesis*. The chiral agent in an asymmetric synthesis may be the substrate, the reagent, or the catalyst. Since biochemical catalysts are chiral, and many reactions occurring in nature are asymmetric syntheses, they are unusally highly specific. In fact most naturally occurring compounds are present as one of their optically active forms rather than as a racemic mixture.

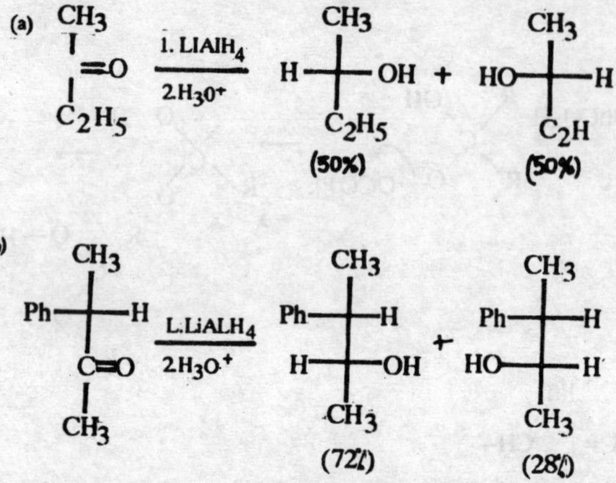

Fig. 29. Examples of symmetric and asymmetric synthesis.

Asymmetric analysis is also observed in the reactions of Grignard reagents with enantiomeric ketones or with esters of α-oxo acids with enantiomeric alcohols. These results have been rationalised in terms of the preferred conformation of the substrate (Cram's rule) and if the absolute configuration of the esterifying alcohol is known then the absolute configuration of the α-hydroxy acid formed in excess can be deduced. This provides a useful method of determining absolute configuration and is detailed in Scheme 11.

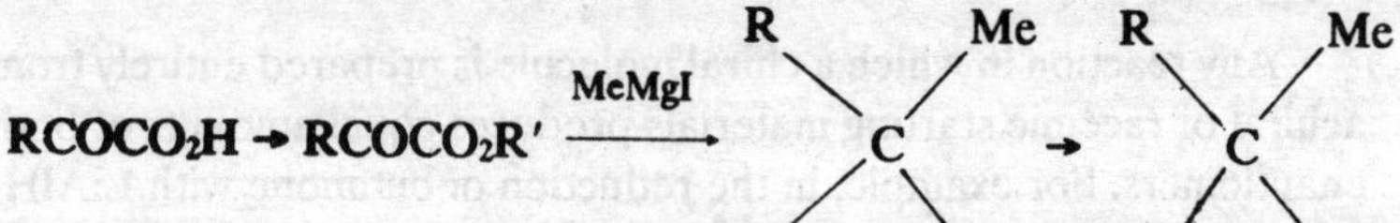

Hydration of an alkene via hydroboration gives a product with a new chiral which will normally be formed as a racemic mixture.

Asymmetric synthesis has been observed when the diborane is replaced by an enantiomeric dialkyborane (R''_2BH) and it is claimed that *cis*-but- 2-ene gives almost pure (–)-butan-2-ol by reaction with the dialkylated borane from an enantiomer of α-pinene.

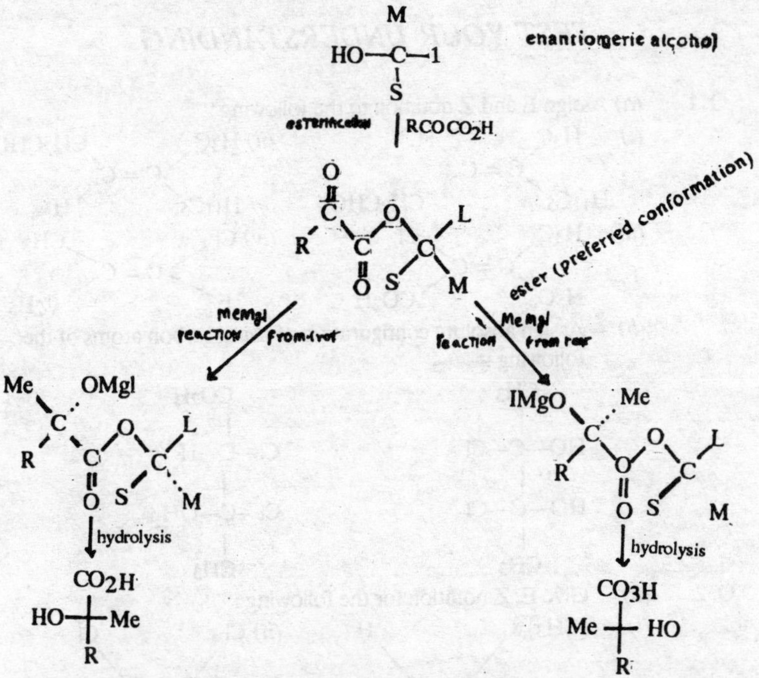

Scheme 11. The enantiomeric alcohol (with large, medium, and small ligands) forms an ester with the preferred conformation shown. The result of reaction from the front and from the back of the carbonyl group is shown on the left and the right respectively. Attack is more likely to occur from the less hindered side (i.e., the front where the ligand is S rather than the back where the ligand is M) so that the hydrolysed acid will contain an excess of that shown on the left. A similar sequence can be used to show that the enantiomeric alcohol would furnish an excess of the enantiomeric hydroxy acid.

Reactions are likely to be influenced by enantiomeric solvents only when solvation of reactants or intermediates is an important feature of the process. The reaction between benzaldehyde and ethylmagnesium bromide gives 1-phenylpropan-1-ol which is optically active when conducted in the presence of (+)-2,3- dimethoxybutane. The enantiomeric ether is part of the Grignard complex which reacts with the aldehyde.

The influence of enantiomeric catalysts has been demonstrated in the preferential formation of an enantiomeric cyanohydrin in the presence of an enantiomeric base such as quinine.

$$PhCH = CHCHO + HCN \xrightarrow[\text{quinine}]{} PhCH = CHCH(OH)CN$$

TEST YOUR UNDERSTANDING

Q. 1. *(a)* Assign E and Z notation to the following :

(i)

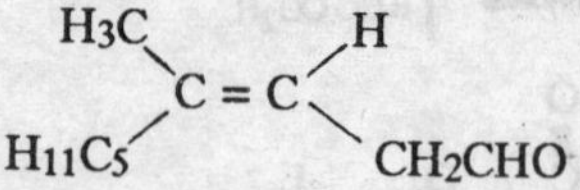

(ii)

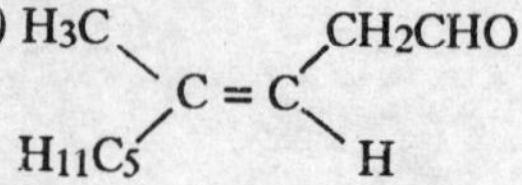

(iii)

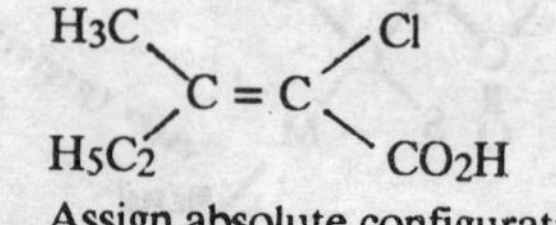

(iv)

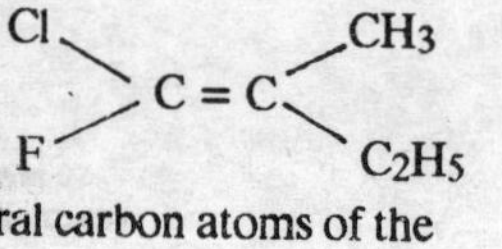

(b) Assign absolute configuration of chiral carbon atoms of the following :

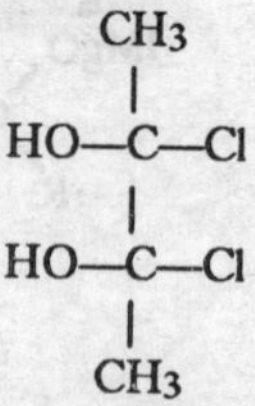

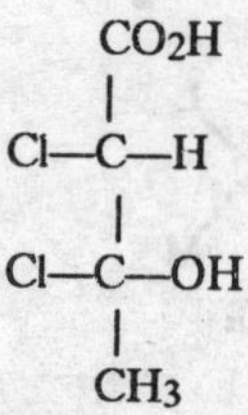

Q. 2. *(a)* Give E, Z notation for the following :

(i)

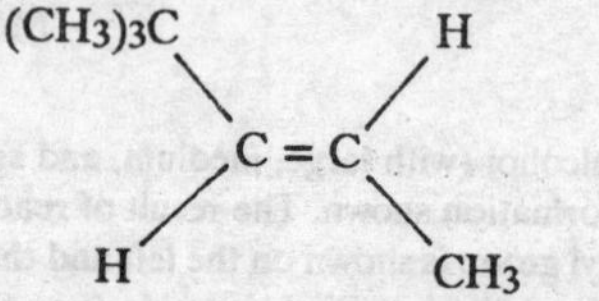

(ii)

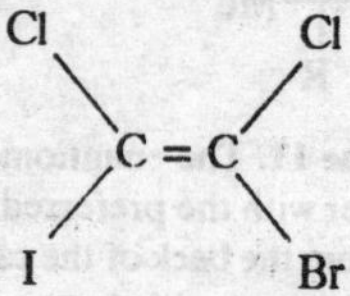

(b) Write R, S notation for the following :

(i)

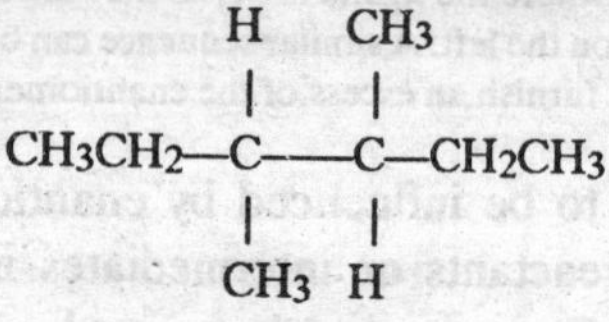

(ii)

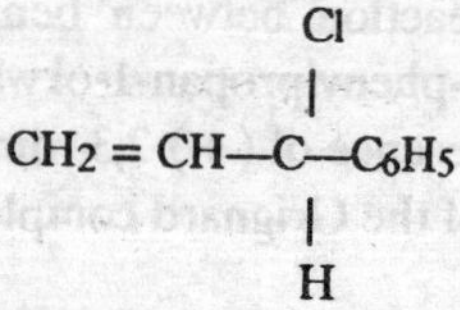

(c) Give Fischer projection formula for

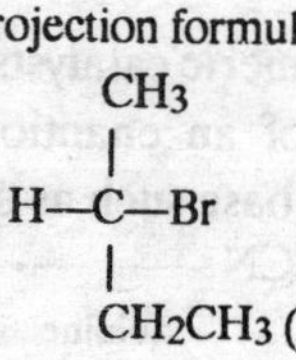

(A)

Which of the following structures are equivalent to (A) :

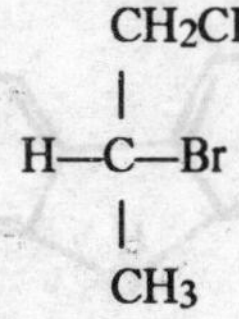

I

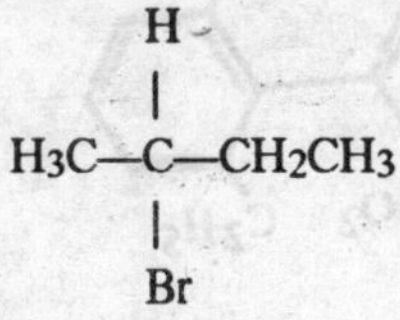

II

Q. 3. Draw the Fischer projection formulae of all the stereoisomers of 2, 3, 4-tribromopentane and point out their R, S specifications, optical activity (where present) and meso nature. *(Delhi B.Sc. Hons., 1988)*

Q. 4. *(a)* Label the following compounds with R or S notation :

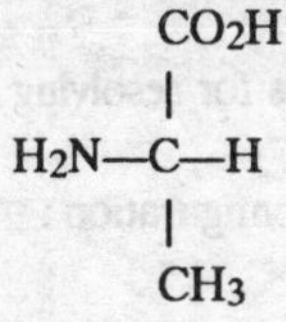

(i)

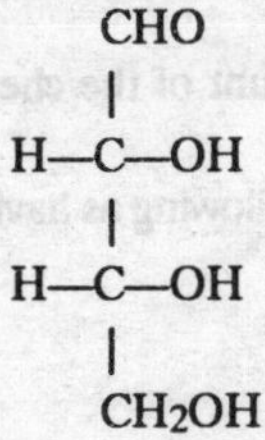

(ii)

(b) Give E or Z designation to the following compounds :

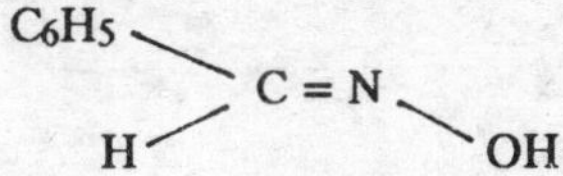

(i)

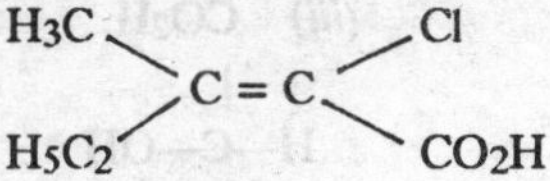

(ii)

Q. 5. *(a)* Write the R or S configuration for the following :

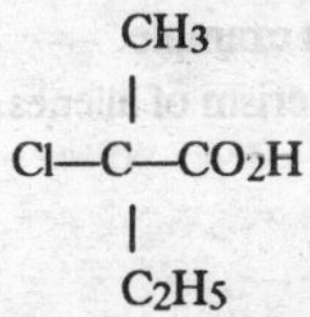

(b) What do you expect when the following substance is dissolved in water ? Explain with the aid of configurations.

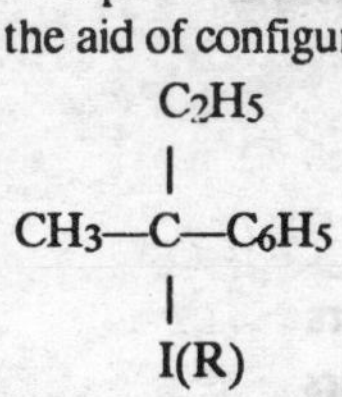

(c) State which of the following substances is resolvable ?

(i) 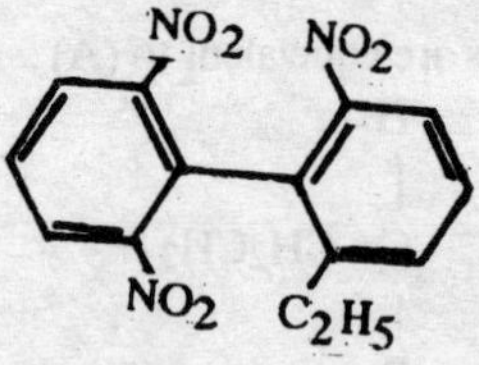

(ii) 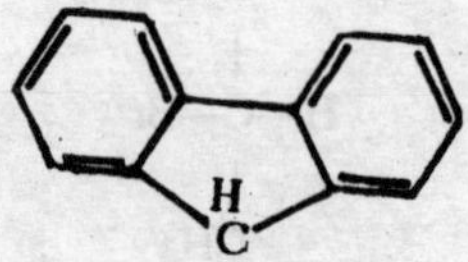

(iii) 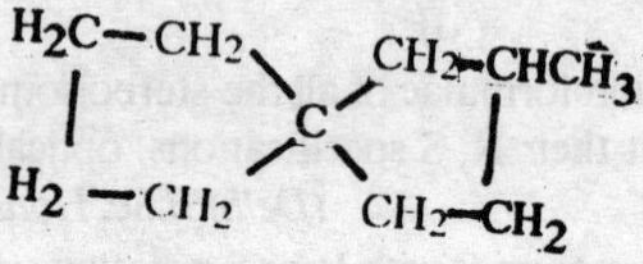

(iv) 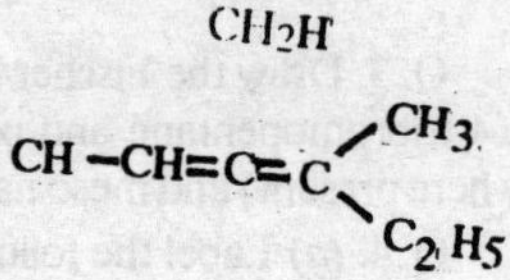

Q. 6. Give an account of the chemical methods for resolving a racemic mixture.

Q. 7. Classify the following as having a R or S configuration :

(i)
```
     Br
     |
  H—C—Cl
     |
     I
```

(ii)
```
     Cl
     |
  H—C—SO3H
     |
     I
```

(iii)
```
     CO2H
     |
  H—C—OH
     |
     CH3
```

Q. 8. What are epimers ? Give an example.

Q. 9. *(a)* Discuss the optical isomerism of allenes.

(b) Write a note on

(i) Asymmetric synthesis.

(ii) R—S notation

(iii) Optical activity of biphenyls.

Q. 10. What are diestereoisomers.

Q. 11. *(a)* Write short notes on

(i) chirality

(ii) Diastereoisomers

(iii) Configurational isomers

(iv) Conformational isomers

(v) R and S notation.

Q. 12. *(a)* Assign E or Z nomenclature to the following compounds giving reasons :

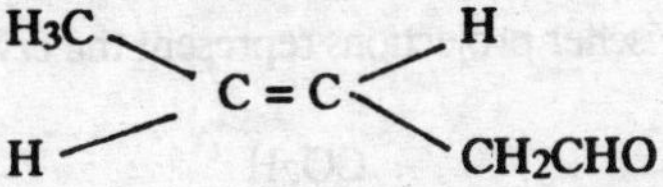

I

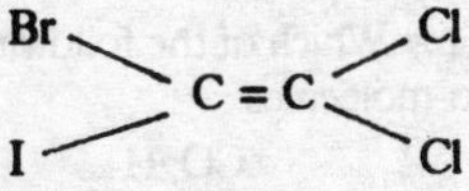

II

(b) Explain with examples the difference between diastereoisomer and an enantiomer.

(c) Differentiate between the conformation and configuration.

Q. 13. *(a)* Draw the Fischer projection formulae and specify as R or S configuration to the enantiomers (if, any) of 3-bromohexane.

(b) Draw the Fischer projection formulae for all the possible stereo-isomers of 2, 3-dichlorobutane. Label the configuration of two relevant carbon atoms by R and S notation in any of them.

(c) Explain the term 1,3-diaxial interaction. Give the Newman projection formulae for the best conformation of methyl- cyclohexane with

(i) methyl group in axial position, and

(ii) methyl group in equatorial position.

Q. 14. *(a)* How many stereoisomers are obtained in the following reaction ? Write their configuration in Fischer projection formula labelling the configuration of each carbon in R and S notation. The compound to begin with has the S configuration.

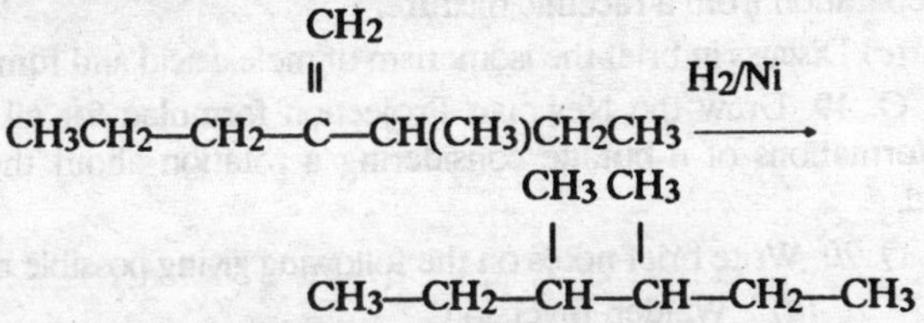

(b) Explain giving reasons why does trans-1,4-ditertiary butyl-cyclo-hexane exist in chair conformation but its geometric isomer cis-1, 4-ditertiarybutyl cyclohexane does not.

Q. 15. *(a)* Differentiate between the configuration and conformation. What are the two important conformations of cyclohexane. Giving reasons, state which one is more stable. Draw axial and equatorial bonds in both the conformations.

(b) Label the asymmetric carbon atoms in the following compounds with R/S notation :

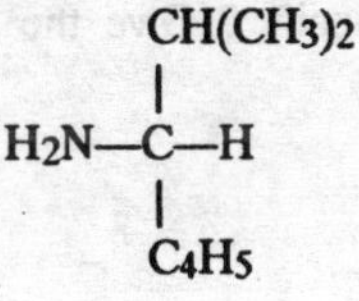

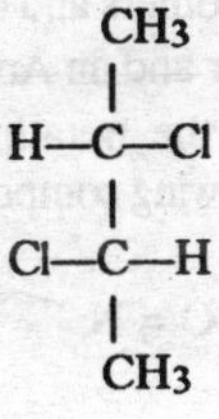

(c) Which of the following Fischer projections represent the erythro- and threo-molecules :

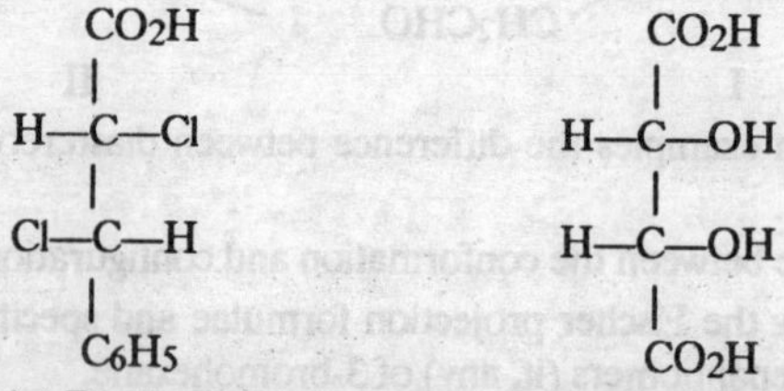

Q. 16. *(a)* Distinguish between (±- and meso tartaric acids.

(b) Give the relationship between observed rotation and specific rotation of an optically active compound.

Q. 17. *(a)* The heat of combustion for methylene of cyclopropane is the same as that in cyclohexane. What should be the C—C—C bond angle in cyclohexane ?

(b) Draw a heat energy diagram for different conformers of cyclohexane. Indicate the most stable and least stable form.

Q. 18. *(a)* What is optical activity ? Discuss the optical isomerism of lactic acid and describe a method for its preparation on an industrial scale.

(b) Explain the existence of optically active lactic acid and the method of its separation from a racemic mixture.

(c) Discuss in brief the isomerism of meleic acid and fumaric acid.

Q. 19. Draw the Newman Projection formulae for all non-equivalent conformations of *n*-butane considering a rotation about the central C—C bond.

Q. 20. Write brief notes on the following giving possible mechanisms :

(a) Walden Inversion,

(b) Asymmetric synthesis,

(c) Racemization.

Q. 21. Explain giving reasons :

(i) Cyclopentane ring is planar although the valency angle for normal pentagon is very near to the tetrahedral carbon angle.

(ii) Boat conformation of cyclohexane is less stable than chair conformation.

Q. 22. Differentiate between the following with the aid of an example :

(a) Enantiomers and diastereoisomers.

(b) Epimer and an Anomer.

Q. 23. What do the notations E and Z signify ? Give the E or Z designation to the following compounds :

(i) 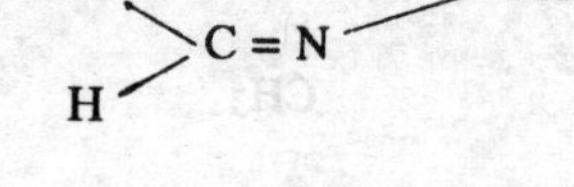

(ii) 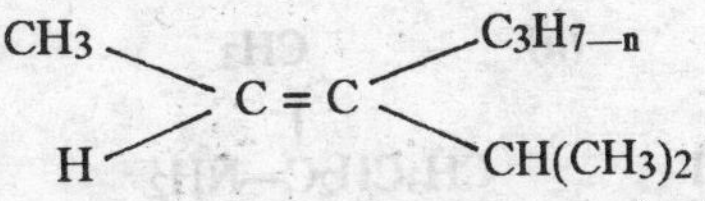

Q. 24. Account for the fact that compounds of the type 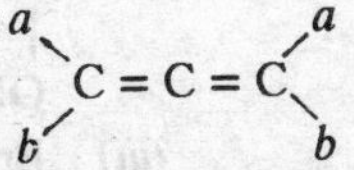 can be resolved into optical isomers.

Q. 25. Discuss Le Bell and van't Hoff theory of optical isomerism with special reference to tartaric acid. How is meso tartaric acid prepared ?

Q. 26. Give an account of the theoretical basis of optical and geometrical isomerism with special reference to *(a)* Lactic acid, *(b)* Tartaric acid and isomers having the formula $C_4H_4O_4$ and $C_4H_5O_5$.

Q. 27. Write structures of the stereo-isomers possible in each or the following :

(a) $H_3C—CHCl\text{-}CHCl—CH_3$.

(b) $H_3C—CHBr—CBr_2—CH_3$.

Q. 28. Write the possible stereoisomers for each of the following :

(a) $CH_2—CH(OH)—CH(OH)—CH_3$.

(b) $HO_2C—CH(CH_3)—CH(Cl)—CO_2H$.

Q. 29. Is it correct to say that a substance that is not asymmetric cannot be resolved ? Explain.

Q. 30. Discuss the optical activity of compounds without asymmetric carbon atom.

Q. 31. Assign R and S designation to each of the following :

(a) 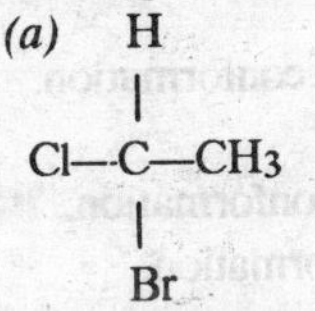

(b) 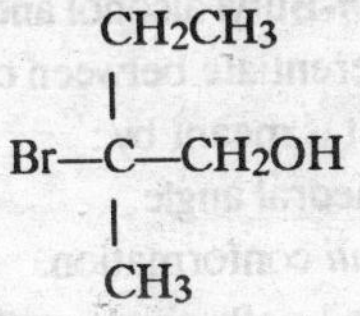

(c) 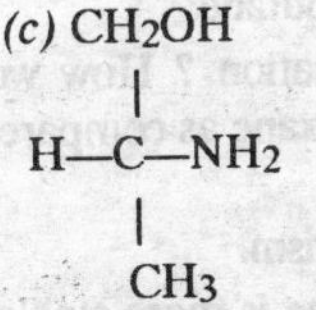

(d) 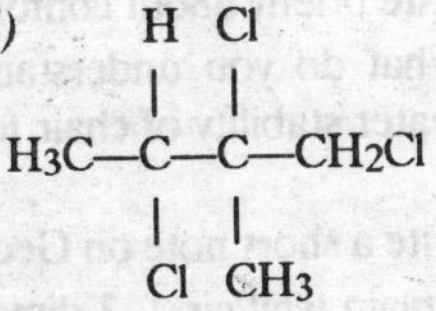

Q. 32. Distinguish between the following terms :

(a) Configuration and conformation.

(b) Absolute and relative configurations.

(c) Observed rotation and specific rotation.

(d) Enantiomer and disastereoisomer.

(e) Meso and racemic forms.

Q. 33. *(a)* Assign R and S notation to the following compounds :

(i) 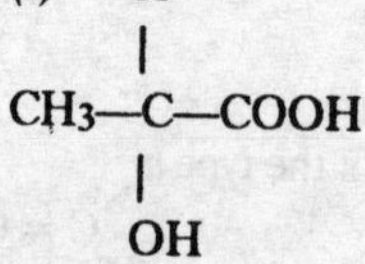

(ii) 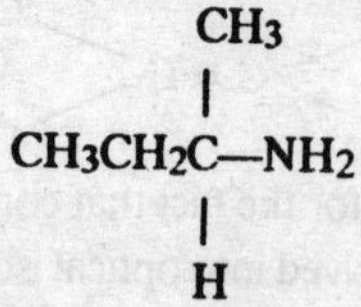

(iii) 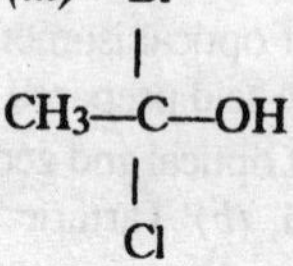

(iii) 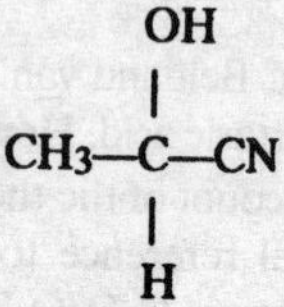

(b) Draw the Newman formula for two cyclohexane conformers.

(c) Write a short note on conformational analysis of *n*-butane.

Q. 34. Write short notes on :

(a) Dee-Ell notation *(b)* E-Z notation.

Q. 35. Write a shcrt note on Stereo isomerism of organic compounds.

Q. 36. Write a short note on Chain and Position Isomerism.

Q. 37. Write briefly on keto-enol tautomerism.

Q. 38. How does resonance differ from tautomerism ?

Q. 39. Mention the specific type of isomerism exhibited by each of the following pairs of compounds ?

(a) 1,2-Dibromoethane and 1,1-dibromoethane.

(b) Vinyl alcohol and acetaldehyde.

(c) *n*-Butyl alcohol and diethyl ether.

Q. 40. Differentiate between configuration and conformation.

Q. 41. What is meant by :

(i) Dihedral angle. *(ii)* Gauche conformation.

(iii) *Anti* conformation. *(iv)* *Syn* conformation.

Q. 42. Write briefly about conformations of *n*-butane.

Q. 43. What do you understand by conformation ? How would you explain the greater stability of chair form of cyclohexane as compared to that of boat form ?

Q. 44. Write a short note on Geometric Isomerism.

Q. 45. Explain why *cis*-1, 3-dimethylcyclohexane is more stable than its trans isomer.

Q. 46. Explain the term 'Polarized light'.

Q. 47. What is the requirement for a compound showing optical acitivity?

Q. 48. Write a short note on optical activity.

Q. 49. Give two examples of optically active compounds.

Q. 50. Lactic acid exhibits optical isomerism whereas propionic acid do not. Explain why ?